Henry B. Spotton

The Commonly Occuring Wild Plants of Canada

and more especially of the province of Ontario - a flora for the use of beginners

Henry B. Spotton

The Commonly Occuring Wild Plants of Canada
and more especially of the province of Ontario - a flora for the use of beginners

ISBN/EAN: 9783337272524

Printed in Europe, USA, Canada, Australia, Japan

Cover: Foto ©berggeist007 / pixelio.de

More available books at **www.hansebooks.com**

W. J. Gage & Co's Educational **Series.**

THE ELEMENTS

OF

STRUCTURAL BOTANY

WITH SPECIAL REFERENCE TO THE STUDY

OF

CANADIAN PLANTS,

TO WHICH IS ADDED

A SELECTION OF EXAMINATION PAPERS.

BY

H. B. SPOTTON, M.A., F.L.S.,

HEAD MASTER OF BARRIE COLLEGIATE INSTITUTE.

THIRD EDITION.

ILLUSTRATED BY THE AUTHOR.

W. J. GAGE AND COMPANY,

TORONTO AND WINNIPEG.

The work, of which the present little volume forms the first part, has been undertaken, at the suggestion of several eminent educationists, to supply a palpable want. The works on Botany, many of them of great excellence, which have found their way into this country, have been prepared with reference to climates differing, in some cases, very widely from our own. They consequently contain accounts of many plants which are entirely foreign to Canada, thus obstructing the search for descriptions of those which happen to be common to our own and other countries ; and, on the other hand, many of our Canadian species are not mentioned at all in some of the Classifications which have been in use. It is believed that the Classification which is to form the second part of this work will be found to contain all the commonly occurring species of the Provinces whose floras it is designed to illustrate, without being burdened with those which are either extremely rare, or which do not occur in Canada at all.

The present Part is designed to teach the Elements of Structural Botany in accordance with a method which is believed to be more rational than that commonly adopted; and it will be found to supply all that is requisite for passing the examinations for Teachers' Certificates of all grades, as well as any others demanding an elementary knowledge of the subject. It contains familiar descriptions of common plants, illustrating the chief variations in plant-structure, with a view to laying a foundation for the intelligent study of Systematic Botany with the aid of the second part; then follow a few lessons on Morphology ; and the

Elements of Vegetable Histology are treated of in as simple and brief a manner as was thought to be consistent with the nature of the subject.

The Schedules, the use of which is very strongly recommended, were devised by the late Professor Henslow, of Cambridge University, to fix the attention of pupils upon the salient points of structure. They will be found invaluable to the teacher as tests of the accuracy of his pupils' knowledge. The cost of striking off a few hundred blanks of each sort would be very trifling, and not worth considering in view of the resulting advantages.

The wood-cuts are from drawings from living specimens, except in two or three instances where assistance was derived from cuts of well-known excellence in standard works on Botany. It need hardly be said that the engravings are not in any sense intended to take the place of the living plants. They are designed chiefly to assist in the examination of the latter, and whilst it is hoped that they may be of service to those who may desire to read the book in the winter season, it is strongly urged upon teachers and students not to be satisfied with them as long as the plants themselves are available.

The works most frequently consulted in the preparation of the text are those of Hooker, Gray, Bentley and Oliver.

Finally, the Authors look for indulgence at the hands of their fellow-teachers, and will be glad to receive suggestions tending to increase the usefulness of the work, and to extend a taste for what must ever be regarded as one of the most refining as well as one of the most practically useful of studies.

September, 1879.

DESCRIPTION OF CUTS.

CONTENTS.

CHAPTER I.

TABLE OF THE COMMON PLANTS EXAMINED, TOGETHER WITH THE FAMILIES TO WHICH THEY BELONG.

Buttercup, Hepatica, Marsh-Marigold.Crowfoot Family.

Shepherd's Purse.. ....Cress Family.

Round-leaved Mallow ....Mallow Family.

Garden Pea ....Pulse Family.

Great Willow-herb ....Evening-Primrose Family.

Sweet-Brier, Crab-Apple...........Rose Family.

Water-Parsnip ....Parsley Family.

Dandelion ....Composite Family.

Catnip ... .Mint Family.

Cucumber ....Gourd Family.

Willow ....Willow Family.

Dog's-tooth Violet, Trillium ....Lily Family.

Indian Turnip, Calla ....Arum Family.

Showy Orchis ....Orchis Family.

Timothy ....Grass Family.

THE ELEMENTS

OF

STRUCTURAL BOTANY.

1. The study of Botany is commonly rendered unattractive to the beginner by the order in which the parts of the subject are presented to him. His patience becomes exhausted by the long interval which must necessarily elapse before he is in a position to do any practical work for himself. In accordance with the usual plan, some months are spent in committing to memory a mass of terms descriptive of the various modifications which the organs of plants undergo; and not until the student has mastered these, and perhaps been initiated into the mysteries of the fibro-vascular system, is he permitted to examine a plant as a whole. In this little work, we purpose, following the example of some recent writers, to reverse this order of things, and at the outset to put into the learner's hands some common plants, and to lead him, by his own examination of these, to a knowledge of their various organs—to

cultivate, in short, not merely his memory, but also, and chiefly, his powers of observation.

It is desirable that the beginner should provide himself with a magnifying glass of moderate power for examining the more minute parts of specimens; a sharp penknife for dissecting; and a couple of fine needles, which he can himself insert in convenient handles, and which will be found of great service in separating delicate parts, and in impaling fine portions for examination with the aid of the lens.

CHAPTER I.

EXAMINATION OF A BUTTERCUP.

2. To begin with, there is no plant quite so suitable as our common Buttercup. This plant, which has conspicuous yellow flowers, may be found growing in almost every moist meadow. Having found one, take up the whole plant, loosening the soil a little, so as to obtain as much of the Root as possible. Wash away the earth adhering to the latter part, and then proceed to examine your specimen. Beginning with the Root, (Fig. 1) the first noticeable

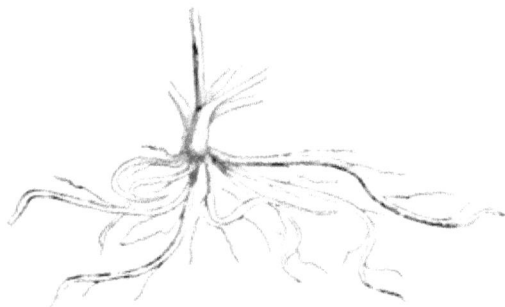

Fig. 1.

thing is that it is not of the same colour as the rest of

the plant. It is nearly white. Then it is not of the same *form* as the part of the plant above ground. It is made up of a number of thread-like parts which spread out in all directions, and if you examine one of these threads through your magnifying glass, you will find that from its surface are given off many finer threads, called *rootlets*. These latter are of great importance to the plant; it is largely by means of their tender extremities, and the parts adjacent to these, that it imbibes the nutritious fluids contained in the soil.

Whilst you are looking at these delicate rootlets, you may perhaps wonder that they should be able to make their way through the soil, but how they do this will be apparent to you if you examine the tip of one of them with a microscope of considerable power. Fig. 2 represents such a tip highly magnified. It is to be observed that the growth of the rootlet does not take place at the very extremity, but immediately behind it. The extreme tip consists of harder and firmer matter than that behind, and is in fact a sort of cap or thimble to protect the growing part underneath. As the rootlet grows, this little thimble is pushed on first through the crevices of the soil, and, as you may suppose, is soon worn away on the outside, but it is as rapidly renewed by the rootlet itself on the inside.

Another difference between the root and the part above ground you will scarcely have failed to discover: the root has no leaves, nor has it any buds.

You may describe the root of the Buttercup as *fibrous*.

3. Let us now look at the Stem. (Fig.3.) It is upright, pretty firm, coloured green, and leaves spring from it at intervals. As there is scarcely any appearance of wood in it, we may describe it as *herbaceous.* At several points along the main stem branches are given off, and you will observe that immediately below the point from which every branch springs there is a leaf on the stem. The angle between the leaf and the stem, on the upper side, is called the *axil* of the leaf (*axilla*, an armpit), and it is a rule to which there are scarcely any exceptions, that branches can only spring from the axils of leaves.

The stem and all the branches of our plant terminate, at their upper extremities, either in flowers or in flower-buds.

Fig. 3

4.   Let us now consider the **Leaves.** A glance will show you that the leaves of this plant are not all alike. Those at the lower end of the stem have long stalks, (Fig. 4) which we shall henceforward speak of as *petioles.* Those a little higher up have petioles too, but they are not

quite so long as the lower ones, and the highest leaves have no petioles at all. They appear to be sitting on the stem, and hence are said to be *sessile*. The lowest leaves of all, as they seem to spring from the root, may be described as *radical*, whilst the higher ones may be called *cauline* (*caulis*, a stem). The broad part of a leaf is its *blade*. In the plant we are now examining, the blades of the leaves are almost divided into distinct pieces, which are called *lobes*, and each of these again is more or less deeply *cut*. Both petioles and blades of our leaves are covered with minute hairs, and so are said to be *hairy*.

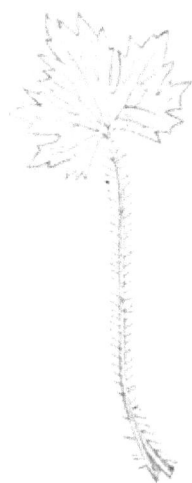

Fig. 4.

Hold up one of the leaves to the light, and you will observe that the veins run through it in all directions, forming a sort of net-work. The leaves are therefore *net-veined*.

The points along the stem from which the leaves arise are called *nodes*, and the portions of stem between the nodes are called *internodes*.

5. Let us next examine the Flowers. Each flower in our plant is at the end either of the stem or of a branch of the stem. The upper portions of the stem and its branches, upon which the flowers are raised, are called the *peduncles* of the flowers.

Take now a flower which has just opened. Beginning at the outside, you will find five little spreading leaves, somewhat yellowish in colour. Each of these is called a *sepal*, and the five together form the calyx of the

Fig. 5

flower. If you look at a flower which is a little older, you will probably not find any sepals. They will have fallen off, and for this reason they are said to be *deciduous*. So, in like manner, the leaves of most of our trees are deciduous, because they fall at the approach of winter. You will find that you can pull off the sepals one at a time, without disturbing those that remain. This shows that they are not connected together. They are therefore said to be *free*, and the calyx is described as *polysepalous*.

Inside the circle of sepals there is another circle of leaves, usually five in number, bright yellow in colour, and much larger than the sepals. Each of them is called a *petal*, and the five together form the corolla of the flower. Observe carefully that each petal is not inserted in front of a sepal, but in front of the space between two sepals. The petals can be removed one at a time like the sepals. They, too, are free, and the corolla is *polypetalous*. If you compare the petals with one another, you will see that they are, as nearly as possible, alike in size and shape. The corolla is therefore *regular*.

6. We have now examined, minutely enough for our present purpose, the calyx and corolla. Though their divisions are not coloured green, like the ordinary leaves of the plant, still, from their general form, you will have no difficulty in accepting the statement that the sepals and petals are in reality *leaves*. It will not be quite so apparent that the parts of the flower which still remain are also only modifications of the same structure. But there is good evidence that this is the case. Let us,

Fig. 6.

however, examine these parts that remain. There is first a large number of little yellow bodies, each at the top of a little thread-like stalk. Each of these bodies, with its stalk, is called a stamen. The little body itself is the *anther*, and the stalk is its *filament*. Your magnifying glass will show you that each anther consists of two oblong sacs, united lengthwise, the filament being a continuation of the line of union. (Fig. 7.)

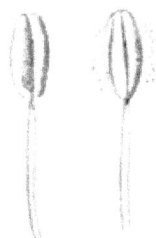

Fig. 7. Fig. 8.

If you look at a stamen of a flower which has been open some time, you will find that each anther-cell has split open along its outer edge, and has thus allowed a fine yellowish dust to escape from it. (Fig. 8.) This dust is called *pollen*. A powerful magnifier will show this pollen to consist of grains having a distinct form.

As the stamens are many in number, and free from each other, they are said to be *polyandrous*.

Fig. 9.

7. On removing the stamens there is still left a little raised mass, (Fig. 9) which with the aid of your needle you will be able to separate into a number of distinct pieces, all exactly alike, and looking something like unripe seeds. Fig. 10 shows one of them very much magnified, and cut through lengthwise. These little bodies, taken separately, are called *carpels*. Taken together, they form the pistil. They are hollow, and

Fig. 10.

each of them contains, as the figure shows, a little grain-like substance attached to the lower end of its cavity. This substance, in its present condition, is the *ovule*, and later on becomes the *seed*.

You will notice that the carpel ends, at the top, in a little bent point, and that the convex edge is more or

Fig. 11.

less rough and moist, so that in flowers whose anthers have burst open, a quantity of pollen will be found sticking there. This rough upper part of the carpel is called the *stigma*. Fig. 11 shows a stigma greatly magnified. In many plants the stigma is raised on a stalk above the ovary. Such a stalk is called a *style*. In the Buttercup the style is so short as to be almost suppressed. When the style is entirely absent the stigma is said to be sessile. The hollow part of the carpel is the *ovary*.

In our plant the pistil is not connected in any way with the calyx, and is consequently said to be *free* or *superior*, and, as the carpels are not united together, the pistil is said to be *apocarpous*.

8. Remove now all the carpels, and there remains nothing but the swollen top of the peduncle. This swollen top is the *receptacle* of the flower. To it. in the case of the Buttercup, all four parts, calyx, corolla, stamens, and pistil, are attached. When a flower has all four of these parts it is said to be *complete*.

9. Let us now return to our statement that the structure of stamens and pistils is only a modification of leaf-structure generally. The stamen looks less like a leaf than any other part of the flower. Fig. 12 will, however, serve to show you the plan upon which the botanist considers a stamen to be formed. The anther corresponds to the leaf-blade, and the filament to the petiole. The two

Fig. 12

cells of the anther correspond to the two

halves of the leaf, and the cells burst open along what answers to the margin of the leaf.

10. In the case of apocarpous pistils, as that of the Buttercup, the botanist considers each carpel to be formed by a leaf-blade doubled lengthwise until the edges meet and unite, thus forming the ovary. Fig. 13 will make this clear.

11. There are many facts which support this theory as to the nature of the different parts of the flower. Suffice it to mention here, that in the white Water-Lily, in which there are several circles of sepals and petals, it is difficult to say where the sepals -end and the petals begin, on account of the gradual change from one set to the other. And

Fig. 13. not only is there a gradual transition from sepals to petals, but there is likewise a similar transition from petals to stamens, some parts occurring, which are neither altogether petals, nor altogether stamens, but a mixture of both, being imperfect petals with imperfect anthers at their summits. We can thus trace ordinary leaf-forms, by gradual changes, to stamens.

We shall, then, distinguish the leaves of plants as foliage-leaves, and flower-leaves, giving the latter name exclusively to the parts which make up the flower, and the former to the ordinary leaves which grow upon the stem and its branches.

12. You are now to try and procure a Buttercup whose flowers, or some of them, have withered away, leaving only the head of carpels on the receptacle. The carpels will have swollen considerably, and will now show themselves much more distinctly than in the

 flower which we have been examining. This is owing to the growth of the ovules, which have now become seeds. Remove one of the carpels, and carefully cut it

Fig. 14.   Fig. 15.

through the middle lengthwise. You will find that the seed almost entirely fills the cavity. (Figs. 14 and 15.)

 This seed consists mainly of a hard substance called *albumen*, enclosed in a thin covering. At the lower end of the albumen is situated a very small body, which is the *embryo*. It is this

Fig. 16.

which develops into a new plant when the seed germinates.

13. We have seen then that our plant consists of several parts :

(1). **The Root.** This penetrates the soil, avoiding the light. It is nearly white, is made up of fibres, from which numbers of much finer fibres are given off, and is entirely destitute of buds and leaves.

(2). **The Stem.** This grows upward, is coloured, bears foliage-leaves at intervals, gives off branches from the axils of these, and bears flowers at its upper end.

(3). **The Leaves.** These are of two sorts : *Foliage-leaves* and *Flower-leaves*. The former are sub-divided into *radical* and *cauline*, and the latter make up the flower, the parts of which are four in number, viz. : calyx corolla, stamens, and pistil.

It is of great importance that you should make yourselves thoroughly familiar with the different parts of the plant, as just described, before going further, and to that end it will be desirable for you to review the present chapter carefully, giving special attention to those parts which were not perfectly plain to you on your first reading.

In the next chapter, we shall give a very brief account of the *uses* of the different parts of the flower. If found too difficult, the study of it may be deferred until further progress has been made in plant examination.

CHAPTER II.

FUNCTIONS OF THE ORGANS OF THE FLOWER.

14. The chief use of the calyx and corolla, or *flora envelops*, as they are collectively called, is *to protect the other parts of the flower*. They enclose the stamens and pistil in the bud, and they usually wither away and disappear shortly after the anthers have shed their pollen, that is, as we shall presently see, as soon as their services as protectors are no longer required.

15. The corollas of flowers are usually bright-coloured, and frequently sweet-scented. There is little doubt that these qualities serve to attract insects, which, in search of honey, visit blossom after blossom, and, bringing their hairy limbs and bodies into contact with the open cells of the anthers, detach and carry away quantities of pollen, some of which is sure to be rubbed off upon the stigmas of other flowers of the same kind, subsequently visited.

16. The essential part of the stamen is the anther, and the purpose of this organ is to produce the pollen, which, as you have already learned, consists of minute grains, having a definite structure. These little grains are usually alike in plants of the same kind. They are

furnished with two coats, the inner one extremely thin, and the outer one much thicker by comparison. The interior of the pollen-grain is filled with liquid matter. When a pollen-grain falls upon the moist stigma it *begins to grow* in a curious manner. (Fig. 17). The inner coat pushes its way through the outer one, at some weak point in the latter, thus forming the beginning of a slender tube. This slowly pene- trates the stigma, and then extends itself down-

Fig. 17. wards through the style, until it comes to the cavity of the ovary. The liquid contents of the pollen grain are carried down through this tube, which remains closed at its lower end, and the body of the grain on the stigma withers away.

The ovary contains an ovule, which is attached by one end to the wall of the ovary. The ovule consists of a kernel, called the *nucleus*, which is usually surrounded by two coats, through both of which there is a minute opening to the nucleus. This opening is called the *micropyle*, and is always to be

Fig. 18. found at that end of the ovule which is not attached to the ovary. (Fig. 18, *m*.)

About the time the anthers discharge their pollen, a little cavity, called the *embryo-sac*, appears inside the nucleus, near the micropyle. The pollen-tube, with its liquid contents, enters the ovary, passes through the micropyle, penetrates the nucleus, and attaches itself to the outer surface of the embryo-sac. Presently the tube becomes empty, and then withers away, and, in the meanwhile, a minute body, which in time developes into the embryo, makes its appearance in the embryo- sac, and from that time the ovule may properly be called a seed.

17. In order that ovules may become seeds, it is always essential that they should be *fertilised* in the manner just described. If we prevent pollen from reaching the stigma—by destroying the stamens, for instance—the ovules simply shrivel up and come to nothing.

Now it is the business of the flower to produce seed, and we have seen that the production of seed depends mainly upon the stamens and the pistil. These organs may consequently be called the *essential organs* of he flower. As the calyx and corolla do not play any *direct* part in the production of seed, but only protect the essential organs, and perhaps attract insects, we can understand how it is that they, as a rule, disappear early. Their work is done when fertilization has been accomplished.

Having noticed thus briefly tne part played by each set of floral organs, we shall now proceed to the examination of two other plants, with a view to comparing their structure with that of the Buttercup.

CHAPTER III.

EXAMINATION OF HEPATICA AND MARSH-MARIGOLD—RESEMBLANCES BETWEEN THEIR FLOWERS AND THAT OF BUTTERCUP.

18. **Hepatica.** You may procure specimens of the Hepatica almost anywhere in rich dry woods, but you will not find it in flower except in spring and early summer. It is very desirable that you should have the plant itself, but for those who are unable to obtain

specimens, the annexed engravings may serve as a substitute.

Beginning then at the root of our new plant, you see that it does not differ in any great measure from that of the Buttercup. It may in like manner be described as *fibrous*.

Fig. 19.

The next point is the stem. You will remember that in the Buttercup the stem is that part of the plant from which the leaves spring. Examining our Hepatica in the light of this fact, and following the petioles of the leaves down to their insertion, we find that they and the roots appear to spring from the same place— that there is, apparently, *no stem*. Plants of this kind are therefore called *acaulescent*, that is, *stemless*, but it must be carefully borne in mind that the absence of the

stem is only apparent. In reality there is a stem, but
it is so short as to be almost indistinguishable.

The leaves of the Hepatica are of course all *radical.*
They will also be found to be *net-reined.*

19. The **Flowers** of the Hepatica are all upon long
peduncles, which, like the leaves, appear to spring from
the root. Naked peduncles of this kind, rising from
the ground or near it, are called *scapes.* The flower-
stalks of the Tulip and the Dandelion furnish other
familiar examples.

Let us now proceed to examine the flower itself.
Just beneath the coloured leaves there are three leaflets, which you will be almost certain to regard, at first
sight, as sepals, forming a calyx. It will not be difficult, however, to convince you that this conclusion
would be incorrect. If, with the aid of your needle,
you turn back these leaflets, you will readily discover,
between them and the coloured portion of
the flower, *a very short bit of stem* (Fig. 20),
the upper end of which is the *receptacle.*
As these leaflets, then, are on the peduncle,
below the receptacle, they cannot be sepals.

Fig. 20.

They are simply small foliage leaves, to which, as they
are found beside the flower, the name *bracts* is given.
Our flower, then, is apparently without a calyx, and in
this respect is different from the Buttercup. The whole
four parts of the flower not being present, it is said to
be *incomplete.*

20. It may be explained here that there is an understanding among botanists, that if the calyx and corolla
are not both present it is always the corolla which is
wanting, and so it happens that the coloured part of
the flower under consideration, though resembling a

corolla, must be regarded as a calyx, and the flower itself, therefore, as *apetalous*.

21. Remove now these coloured sepals, and what is left of the flower very much resembles what was left of our Buttercup, after the removal of the calyx and corolla. The stamens are very numerous, and are inserted on the receptacle. The carpels are also numerous, (Fig. 21) are inserted on the receptacle, and are free from each other (*apocarpous*). And if you

Fig. 21.   Fig. 22.

examine one of the carpels (Fig. 22) you will find that it contains a single ovule. The flower, in short, so much resembles that of the Buttercup that you will be prepared to learn that the two belong to the same Order or Family of plants, and you will do well to observe and remember such resemblances as have just been brought to your notice, when you set out to examine plants for yourselves, because it is only in this way, and by slow steps, that you can acquire a satisfactory knowledge of the reasons which lie at the foundation of the classification of plants.

22. Marsh-Marigold. This plant grows in wet places almost everywhere, and is in flower in early summer.

Note the entire absence of hairs on the surface of the plant. It is therefore *glabrous*.

The root, like that of the Buttercup and of the Hepatica, is *fibrous*.

The stem is hollow and furrowed.

The foliage-leaves are of two kinds, as in the Buttercup. The radical leaves spring from the base of the stem, whilst the higher ones are cauline. The leaves

are not lobed, as in the other two plants, but are indented on the edge. They are also net-veined.

23. Coming to the flower (Fig. 23)
we find a circle, or whorl, of bright
yellow leaves, looking a good deal
like the petals of the Buttercup, but
you will look in vain for the corresponding sepals. In this case there
is no whorl of bracts to mislead you.
Are we to say, then, that there is no
calyx ? If we adhere to the understanding mentioned when describing
the Hepatica, we must suppose the
corolla to be wanting, and then the
bright yellow leaves of our plant will

Fig. 23.

be the *sepals*, and will together constitute the calyx.
As to the number of the sepals, you will find, as in the
Hepatica, some variation. Whilst the normal number
is five, some flowers will be found to have as many as
nine.

24. The stamens are next to be examined, but you
should first satisfy yourselves as to whether the calyx
is polysepalous or otherwise, and whether it is free from
the other floral leaves or not. If your examination
be properly made, it will show you that the calyx
is free and polysepalous.

The stamens are very much like those of the Buttercup and Hepatica. They are numerous, they have both
anthers and filaments, and they shed their pollen
through slits on the outer edges of the anthers. They
are all separate from each other (polyandrous) and are
all inserted on the receptacle. On this latter account
they are said to be *hypogynous*.

25. Remove the stamens, and you have left, as before, a head of carpels (Fig. 24). Examine one : there is the lower broad part, which you recognize as the *ovary*, the very short style, and the sticky stigma. To all appearance the carpels are pretty much the same as those of the two plants already examined. It will not do, however, to trust altogether to appearances in this case. Cut open a carpel and you find that, instead of a single ovule at the bottom of the ovary, there are several ovules in a row along that edge of the ovary which is turned towards the centre of the flower.

Fig. 24.

The ovary is, in fact, a *pod*, and, when the seeds ripen, splits open along its inner edge. If you can find one which has split in this way, you can hardly fail to be struck with the resemblance which it bears to a common leaf. (Fig. 25.)

Fig. 25.

On the whole the resemblance between the structure of the Marsh-marigold and that of the Hepatica and Buttercup is sufficiently great to justify us in placing it in the same family with them.

26. Having now made yourselves familiar with the different parts of these three plants, you are to write out a tabular description of them according to the following form ; and, in like manner, whenever you examine a new plant, do not consider your work done until you have written out such a description of it.

In the form the term *cohesion* relates to the union of *like* parts ; for example, of sepals with sepals, or petals with petals ; while the term *adhesion* relates to the union of *unlike* parts ; for example, of stamens with corolla, or ovary with calyx. Neither cohesion nor adhesion takes place in any of the three flowers we have examined, and accordingly, under these headings in our schedule we write down the terms polysepalous, polypetalous, &c., to indicate this fact.

The symbol ∞ means "indefinite," or "numerous," and may be used when the parts of any organ exceed ten in number.

BUTTERCUP.

ORGAN OR PART OF FLOWER.	NO.	COHESION.	ADHESION.	REMARKS.
Calyx.		Polysepalous.	Inferior.	
Sepals.	5			
Corolla.		Polypetalous. Regular.	Hypogynous.	
Petals.	5			
Stamens.	∞	Polyandrous.	Hypogynous.	
Filaments.				
Anthers.				
Pistil.		Apocarpous.		
Carpels.	∞			
Ovary.			Superior.	

HEPATICA.

ORGAN.	NO.	COHESION.	ADHESION.	REMARKS.
C .lyx. *Sepals.*	7-12	Polysepalous.	Inferior.	Coloured like a Corolla.
Calyx. *Petals.*				Wanting.
Stamens. *Filaments. Anthers.*	x	Polyandrous.	Hypogynous.	
Pistil. *Carpels. Ovary.*	x	Apocarpous.	Superior.	

MARSH-MARIGOLD.

ORGAN.	NO.	COHESION.	ADHESION.	REMARKS.
Calyx. *Sepals*	5-9	Polysepalous.	Inferior.	Coloured like Corolla.
Corolla *Petals.*				Wanting.
Stamens. *Filaments. Anthers.*	∝	Polyandrous.	Hypogynous.	
Pistil. *Carpels. Ovary.*	∝	Apocarpous.	Superior.	Carpels contain several seeds.

CHAPTER IV.

EXAMINATION OF OTHER COMMON PLANTS WITH HYPOGYNOUS
STAMENS. SHEPHERD'S PURSE. ROUND-LEAVED MALLOW.

27. We shall now proceed to examine some plants. the flowers of which exhibit, in their structure, important variations from the Buttercup, Hepatica, and Marsh-Marigold.

Shepherd's Purse. This plant, (Fig. 26). is one of the commonest of weeds. As in the Buttercup, the foliage-leaves are of two kinds, radical and cauline, the former being in a cluster around the base of the stem. The cauline leaves are all sessile, and each of them, at its base, projects backward on each side of the stem, so that the leaf somewhat resembles the head of an arrow. Such leaves are, in fact, said to be *sagittate*, or arrow-shaped. The flowers grow in a cluster at the top of the stem, and, as the season advances, the peduncle gradually elongates, until, at the close of the summer, it forms perhaps half of the entire length of the stem. You will observe, in this plant, that each separate flower is raised on a little stalk of its own. Each of these little stalks is a *pedicel*, and when pedicels are present, the term peduncle is applied to the portion of stem which supports the whole cluster.

28. The flowers, (Fig. 27), are rather small, and so will require more than ordinary care in their examination. The calyx is polysepalous, and of four sepals. The corolla is polypetalous, and of four petals. The stamens, (Fig. 28), are six in number, and if you examine them attentively, you will see that two of them are shorter than the other four. The stamens are consequently said to be *tetradynamous*. But if there had been only *four* stamens, in two sets of two each, they would have been called *didynamous*. The stamens are inserted on the receptacle (hypogynous). The pistil is separate from the other parts of the flower (superior).

Fig. 27.

Fig. 28.

29. To examine the ovary, it will be better to select a ripening pistil from the lower part of the peduncle.

It is a flat body, shaped something like a heart, (Fig 29) and having the short style in the notch. A ridge divides it lengthwise on each side. Carefully cut or pull away the lobes, and this ridge will remain, presenting now the appearance of a narrow loop, with a very thin membranous partition stretched across it. Around the edge, on both sides of the partition, seeds are suspended from slender stalks. (Fig. 30). There are, then, *two carpels* united together, and the pistil is, therefore, *syncarpous.*

Fig. 29.   Fig. 30

The peculiar pistil of this flower should be carefully noticed, as it is the leading character of a whole group of plants. When you meet with such a pistil, you may be pretty certain that the plant to which it belongs is a member of the *Cress* or *Crucifer* family, so called from the four petals sometimes spreading out like the arms of a cross. We shall find, however, that there are cross-shaped corollas belonging to plants of other groups.

SHEPHERD'S PURSE.

ORGAN.	No.	COHESION.	ADHESION.	REMARKS
Calyx. Sepals.	4	Polysepalous.	Inferior.	
Corolla. Petals.	4	Polypetalous.	Hypogynous.	
Stamens. Filaments. Anthers.	6	Tetradynamous.	Hypogynous.	
Pistil. Carpels. Ovary.	2	Syncarpous.	Superior.	The two cells of the ovary separated by a thin partition.

30. Mallow. The round-leaved Mallow (Fig. 31) grows along every way side, and is a very common weed in cultivated grounds. Pro- cure, if possible, a plant which has ripened its seeds, as well as one in flower. The root of this plant is of a different kind from those of the three plants first examined. It consists of a stout tapering part, descending deep into the soil, from the surface of which fibres are given off irregularly. A stout root of this kind is called a *tap-root*. The Carrot is another example.

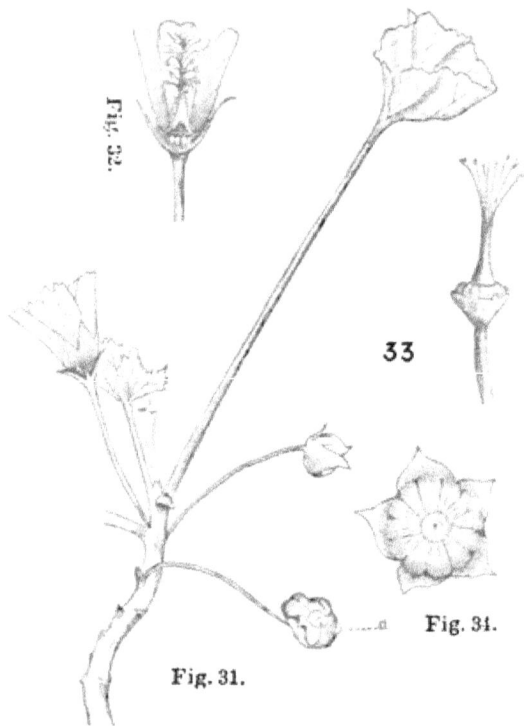

Fig. 32.

33

Fig. 34.

Fig. 31.

31. The leaves are long-petioled, net-veined and in- dented on the edges. On each side of the petiole, at its junction with the stem, you will observe a little leaf- like attachment, to which the name *stipule* is given. The presence or absence of stipules is a point of some importance in plant-structure, and you will do well to notice it in your examinations. You have now made yourselves acquainted with all the parts that any leaf has, viz., *blade*, *petiole* and *stipules*.

32. Coming to the flower, observe first that the parts of the calyx are not entirely separate, as in the flowers you have already examined. For about half their length they are united together so as to form a cup. The upper half of each sepal, however, is perfectly distinct, and forms a *tooth* of the calyx ; and the fact that there are five of these teeth shows us unmistakably that the calyx is made up of five sepals. We therefore speak of it as a *gamosepalous* calyx, to indicate that the parts of it are coherent.

As the calyx does not fall away when the other parts of the flower disappear, it is said to be *persistent*. Fig. 31, *a*, shows a persistent calyx.

33. At the base of the calyx there are three minute leaf-like teeth, looking almost like an outer calyx. A circle of bracts of this kind is called an *involucre*. The three bracts under the flower of the Hepatica also constitute an involucre. As the bracts in the Mallow grow on the calyx, some botanists speak of them as an *epicalyx*.

The corolla consists of five petals, separate from each other, but united with the stamens at their base.

34. The stamens are numerous, and as their filaments are united to form a tube they are said to be *monadelphous*. This tube *springs from the receptacle*, and the stamens are therefore *hypogynous*. Fig. 32 will help you to an understanding of the relation between the petals and stamens.

Having removed the petals, split the tube of the stamens with the point of your needle. A little care will then enable you to remove the stamens without injuring the pistil. The latter organ will then be found to consist of a ring of coherent carpels, a rather stout style, and numerous long stigmas. (Fig. 33.) If you take the trouble to count the carpels and the stigmas, you will find the numbers to correspond. As the seeds ripen the carpels separate from each other. (Fig. 34.)

MALLOW.

ORGAN.	No.	COHESION.	ADHESION.	REMARKS.
Calyx. Sepals.	5	Gamosepsa-lous.	Inferior.	Three bracts growing on the Calyx.
Corolla. Petals.	5	Polypetalous.	Hypogynous.	
Stamens. Filaments. Anthers.	∝	Monadelphous United in a ring. One-celled.	Hypogynous.	
Pistil. Carpels. Ovary.	∞	Syncarpous.	Superior.	Carpels as many as the stigmas.

CHAPTER V.

EXAMINATION OF **COMMON PLANTS** WITH PERIGYNOUS STAMENS —GARDEN **PEA. GREAT WILLOW-HERB, SWEET-BRIER,** CRAB-APPLE.

35. Garden Pea. In the flower of this plant, the calyx is constructed on the same plan as in the Mallow. There are five sepals, coherent below, and spreading out into distinct teeth above (Fig. 35). The calyx is there-fore gamosepalous.

Examine next the **form of the** corolla (Fig. 36). One **difference** between **this** corolla and those of the previous plants **will strike** you at once. In the flowers

of the latter you will remember that each petal
was precisely like its
fellows in size and
shape, and we there-
fore spoke of the corolla
as *regular*. In the Pea,
on the other hand, one
of the petals is large,
broad, and open, whilst

Fig. 36.

Fig. 35.

Fig. 38.

Fig. 39.

Fig. 37.

two smaller ones, in the front of the flower, are
united into a kind of hood. We shall speak of this
corolla, then, and all others in which the petals are
unlike each other in size or shape, as *irregular*.

As the Pea blossom bears some resemblance to a
butterfly, it is said to be *papilionaceous*.

36. Remove now the calyx-teeth and the petals,
being very careful not to injure the stamens and the
pistil, enveloped by those two which form the hood.
Count the stamens, and notice their form (Fig. 37).
You will find ten, one by itself, and the other nine with
the lower halves of their filaments joined together, or
coherent. When stamens occur in this way, in two
distinct groups, they are said to be *diadelphous*; if in
three groups, they would be *triadelphous*; if in several
groups, *polyadelphous*. In the Mallow, you will remember, they are united into *one* group, and therefore we
described them as *monadelphous*.

You will perhaps be a little puzzled in trying to
determine to what part of the flower the stamens are
attached. If you look closely, however, you will see
that the attachment, or *insertion*, is not quite the same
as in the Buttercup and the other flowers examined.

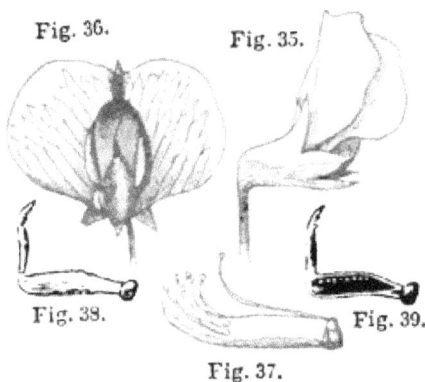

In the present instance, they are inserted upon the lower part of the calyx, and so they are described as *perigynous,* a term meaning "around the pistil."

37. But the pistil (Figs. 38, 39) is not attached to the calyx. It is *free,* or *superior.* If you cut the ovary across, you will observe there is but one cell, and if you examine the stigma, you will find that it shows no sign of division. You may therefore be certain that the pistil is a single carpel.

You are now prepared to fill up the schedule descriptive of this flower.

GARDEN PEA.

ORGAN.	No.	COHESION.	ADHESION.	REMARKS.
Calyx. Sepals.	5	Gamosepalous.	Inferior.	
Corolla. Petals.	5	Polypetalous. Irregular.	Hypogynous.	
Stamens. Filaments. Anthers.	10	Diadelphous.	Perigynous.	
Pistil. Carpels. Ovary.	. 1	Apocarpous.	Superior.	

38. The beginner will be very likely to think, from its appearance, that the largest of the petals is made up of two coherent ones, but the following considerations show clearly that this is not the case. In the Buttercup, and other flowers in which the number of sepals and petals is the same, the petals do not stand

before the sepals, but before the spaces between them. In the Pea-blossom this rule holds good if the large petal is considered as one, but not otherwise. Again, the veining of this petal is similar to that of a common leaf, there being a central rib from which the veins spring on each side ; and lastly, there are some flowers of the Pea kind—Cassia, for example—in which this particular petal is of nearly the same size and shape as the other four.

39. **Great Willow-herb.** This plant is extremely common in low grounds and newly cleared land, and you may easily recognize it by its tall stem and bright purple flowers.

Observe the position of the flowers. In the three plants first examined we found the flowers at the end of the stem. In the Willow-herb, as in the Mallow, they spring from the sides of the stem, and immediately below the point from which each flower springs you will find a small leaf or bract (Fig. 40.) Flowers which arise from the axils of bracts are said to be *axillary*, whilst those which are at the ends of stems are called *terminal*, and you may remember that flowers can only be produced in the axils of leaves and at the ends of stems and branches.

Fig. 40.

40. Coming to the flower itself, direct your attention, first of all, to the position of the ovary. You will find it apparently under the flower, in the form of a tube tinged with purple. It is not in reality under the flower, because its purplish covering is the calyx, or

more accurately the *calyx-tube*, which adheres to the whole surface of the ovary, and expands above into four long teeth. The ovary therefore is *inferior*, and the calyx of course *superior*, in this flower. As the sepals unite below to form the tube the calyx is gamosepalous.

The corolla consists of four petals, free from each other, and is consequently polypetalous. It is also regular, the petals being alike in size and shape. Each petal is narrowed at the base into what is called the *claw* of the petal, the broad part, as in the ordinary foliage-leaf, being the *blade*.

The stamens are eight in number (octandrous), four short and four long, and are attached to the calyx (perigynous).

41. The pistil has its three parts, ovary, style, and stigma, very distinctly marked. The stigma consists of four long lobes, which curl outwards after the flower opens. The style is long and slender. The examination of the ovary requires much care. You will get the best idea of its structure by taking one which has just burst open, and begun to discharge its seeds (Fig. 41). The outside will then be seen to consist of four pieces *(valves)*, whilst the centre is occupied by a slender four-winged column, (Fig. 42), in the grooves of which the seeds are compactly arranged. The pistil thus consists

Fig. 42.

Fig. 41.

of four carpels united together, and is therefore *syncarpous*. Every seed is furnished with a tuft of silky hairs, which greatly facilitates its transportation by the wind.

42. The Willow-herb furnishes an excellent example of what is called *symmetry*. We have seen that the calyx and corolla are each made up of four parts; the stamens are in two sets of four each; the stigma is four-lobed, and the ovary has four seed-cells. A flower is *symmetrical* when each set of floral leaves contains either the same number of parts or a *multiple* of the same number.

Observe that the leaves of our plant are net-veined.

The schedule will be filled up as follows :

GREAT WILLOW-HERB.

Organ	No.	Cohesion.	Adhesion.	Remarks.
Calyx. Sepals.	4	Gamosepalous.	Superior.	
Corolla. Petals.	4	Polypetalous.	Perigynous.	
Stamens. Filaments. Anthers.	8	Octandrous.	Perigynous.	Four short and four long.
Pistil. Carpels. Ovary.	4	Syncarpous.	Inferior.	Seeds provided with tufts of hair

43. Sweet Brier. As in the flower just examined, the

sepals of Sweet-Brier are not entirely distinct; their lower halves cohere to form a tube, and the calyx is therefore gamosepalous.

The corolla consists of five separate petals of the same size and shape, and is therefore both regular and polypetalous.

Fig. 43.

The stamens are very numerous, and separate from each other. As in the Pea and the Willow-herb, so in this flower they will be found to be attached to the calyx. They are, therefore, *perigynous.*

44. To understand the construction of the pistil, you must make a vertical section through the roundish green mass which you will find on the under side of the flower. You will then have presented to you some such appearance as that in Fig. 44. The green mass, you will observe, is hollow. Its outer covering is simply the continuation of the calyx-tube. *The lining of this calyx-tube is the receptacle of the flower;* to it are attached the separate carpels which together

Fig. 44.

constitute the pistil (Fig. 45), just as the carpels of the Buttercup are attached to the *raised* receptacle of that flower.

Fig 45.

We must remind you again that whenever the ovary is enclosed in the calyx-tube, and the calyx appears to spring from the summit of the ovary, the latter is said to be *inferior*, and the former *superior*.

SWEET-BRIER.

ORGAN.	NO.	COHESION.	ADHESION.	REMARKS.
Calyx. Sepals.	5	Gamosepalous	Superior.	
Corolla Petals.	5	Polypetalous.	Perigynous.	
Stamens.	∞	Polyandrous.	Perigynous.	
Pistil. Carpels.	∞	Apocarpous.	Inferior.	The hollow receptacle lines the calyx-tube

45, **Crab-Apple.** The flower of the Crab-Apple (Fig. 46), is in most respects, like that of Sweet-Brier. The calyx is gamosepalous, its parts being united below into a tube. The corolla is of five separ-

Fig 46.

Fig. 47.

ate petals. The stamens are numerous and are inserted on the calyx.

The structure of the pistil (Figs. 47, 48), however, is somewhat different. On making a cross-section through the young apple, five cells containing the unripe seeds are seen radiating from the centre. These seed-vessels are imbedded in a fleshy mass, the outer limit of which is marked by a circle of green dots, and outside these dots is the flesh which constitutes the eatable part of the apple. The inner mass, which encloses the core, belongs to the receptacle, whilst the outer edible portion is the enlarged calyx. At the end opposite the stem will be found the persistent calyx-teeth. We have in this flower, therefore, a *syncarpous* pistil of five carpels, instead of an apocarpous one, as in Sweet-Brier.

Fig. 48.

CRAB-APPLE.

ORGAN.	NO.	COHESION.	ADHESION.	REMARKS.
Calyx. Sepals.	5	Gamosepalous	Superior.	
Corolla. Petals.	5	Polypetalous.	Perigynous.	
Stamens.	*x*	Polyandrous.	Perigynous.	
Pistil. Carpels.	5	Syncarpous.	Inferior.	Fruit consists chiefly of a fleshy enlargement of the Calyx-tube.

CHAPTER VI.

EXAMINATION OF A PLANT WITH EPIGYNOUS STAMENS—
WATER PARSNIP.

16. **Water-Parsnip.** This is a common swamp plant in Canada ; but if any difficulty be experienced in procuring specimens the flower of the common Carrot or Parsnip may be substituted for it, all these plants being closely related, and differing but slightly in the structure of their flowers.

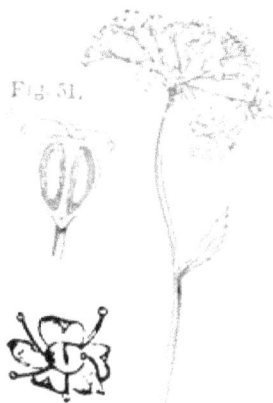

Notice first the peculiar appearance of the flower cluster. (Fig. 49.) There are several pedicels, nearly of the same length, radiating from the end of the peduncle, and from the end of each pedicel radiate in like manner a number of smaller ones, each with a flower at its extremity. Such a cluster is known as an umbel. If, as in the present case, there are groups of secondary pedicels, the umbel is compound. As the flowers are very small we shall be obliged to use the lens all through the examination. Even with its aid you will have a little difficulty in making out the calyx, the tube of which, in this flower, adheres to the surface of the ovary, as in Willow-herb, and is reduced above to a mere rim or border, of **five** minute teeth. The petals are five in number, and **free from** each other. Observe that each of them is incurred at its extremity. (Fig. 50.) They are inserted on a disk which crowns the

Fig. 51.

Fig. 50.   Fig. 49.

ovary, as are also the five stamens, which are hence said to be *epigynous*. In the centre of the flower are two short styles projecting above the disk, and a vertical section through the ovary (Fig. 51) shows it to be two-celled, with a single seed suspended from the top of each cell.

WATER-PARSNIP.

ORGAN.	NO.	COHESION.	ADHESION.	REMARKS.
Calyx. Sepals.	5	Gamosepalous.	**Superior.**	Calyx-teeth almost obsolete.
Corolla. Petals	5	**Polypetalous.**	Epigynous	**Petals** incurved.
Stamens.	5	Pentandrous.	Epigynous.	
Pistil. Carpels.	2	Syncarpous.	Inferior.	

CHAPTER VII.

EXAMINATION OF COMMON **PLANTS WITH EPIPETALOUS** STAMENS —DANDELION CATNIP.

47. Dandelion. The examination of this flower will be somewhat more difficult than that of any we have yet undertaken.

Provide yourselves with specimens in flower and in seed.

The root of the plant, like that of the Mallow, is a tap-root.

The stem is almost suppressed, and, as in the case of the Hepatica, the leaves are all radical. They are also net-veined.

The flowers are raised on scapes, which are hollow. At first sight the flower appears to have a calyx of many sepals, and a corolla of many petals. Both of these appearances, however, are contrary to facts. With a sharp knife cut the flower through the middle from top to bottom. (Fig. 52.) It will then appear that the flower or rather *flower-head*, is made up of a large number of distinct pieces. With the point of your needle detach one of these pieces. At the lower end of it you have a small body resembling an un-ripe seed. (Fig. 53.) It is, in fact, an ovary. Just above this there is a short bit of stalk, sur-mounted by a circle of silky hairs, and above this a yellow tube with one side greatly prolonged. This yellow tube is a corolla, and a close examina-tion of the extremity of its long side will show the existence of five minute points, or teeth, from which we infer that the tube is made up of five coher-ent petals. As the corolla is on the ovary it is said to be *Epigynous*.

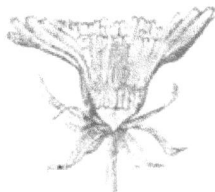

Fig. 52.

Fig. 53.

Out of the corolla protrudes the long style, divi-ded at its summit into two stigmas.

To discern the stamens will require the greatest nicety of observation. Fig. 54 will help you in your task. The stamens are five in number. They are inserted on the tube of the corolla (epipetalous) and their anthers cohere (Fig. 55) and form a ring about the style. When the anthers are united

Fig. 54. in this way, the stamens are said to be *syngenesious*.

48. It appears, then, that the Dandelion, instead of being a single flower, is in reality a compound of a great many flowers upon a common receptacle, and what seemed at first to be a calyx is, in reality, an *involucre*, made up of many bracts.

Fig. 55.

But have the single flowers, or *florets*, as they are properly called, no calyx ? The theory is that they have one, but that it is adherent to the surface of the ovary, and that the tuft of silky hairs which we noticed is a prolongation of it.

Now turn to your specimen having the seeds ready to blow away. The seeds are all single ; the little bit of stalk at the top has grown into a long slender thread, and the tuft of hairs has spread out like the rays of an umbrella (Fig. 56). But though the seeds are invariably single, it is inferred from the two-lobed stigma that there are *two carpels*.

49. Flowers constructed on the plan of the Dandelion are called *composite* flowers. A very large number of our common plants

Fig. 56.

have flowers of this kind. The May-weed, which abounds in waste places everywhere, the Thistle, and the Ox-Eye Daisy are examples.

DANDELION.

ORGAN.	No.	COHESION.	ADHESION.	REMARKS.
Calyx. *Sepals.*	5	Gamosepalous.	Superior.	The number of sepals is *inferred from analogy* to be five.
Corolla. *Petals.*	5	Gamopetalous.	Epigynous.	
Stamens.	5	Syngenesious.	Epipetalous.	
Pistil. *Carpels.*	2	Syncarpous.	Inferior.	Number of carpels inferred from number of stigmas.

50. Catnip. Note carefully the appearance of the stem. It is *square.*

The flowers are in axillary clusters. The calyx is a tube (Fig. 57) terminating in five sharp teeth, and you may observe that the tube is a little longer on the upper side (that is, the side *towards* the stem) than on the lower. The corolla is some-  what peculiar. It has somewhat the appearance of a wide open mouth, and is known as a *labiate* or two-lipped corolla. The upper lip is erect, and notched at the apex. The lower lip spreads outward, and Fig. 57. consists of a large central lobe and two small lateral ones. Altogether, therefore, there are *five* lobes consti- tuting the gamopetalous corolla. Pull out the corolla, and with the point of your needle split its tube in front. On laying it open, the stamens will be found to be in- serted upon it (epipetalous). They are four in number,

 two of them shorter than the other two. Hence they are described as *didynamous.* The anthers are peculiar in not having their lobes parallel (Fig. 58), these being wide apart at the base, in consequence of the expansion of the *connective*, the name given to that part of
Fig. 58. the anther which unites its two lobes or cells.

The pistil consists of a two-lobed stigma, a long style, and an ovary which seems at first as if made up of four distinct carpels (Fig. 59).  But the single style and the two-lobed stigma will warn you against this supposition. The ovary really consists of two carpels, each of two deep lobes, and, as the seeds ripen, these lobes form four little nutlets (Fig. 60), each contain- Fig. 59. ing a single seed.

51. The group of plants to which Catnip belongs is easily distinguished by the square stem, irregular corolla, and four stamens.

Fig. 60.

CATNIP.

ORGAN.	NO.	COHESION.	ADHESION.	REMARKS.
Calyx. Sepals.	5	Gamosepalous.	Inferior.	
Corolla. Petals.	5	Gamopetalous	Hypogynous.	Two-lipped. Upper lip of two, and lower of three lobes.
Stamens. Anthers.	4	Didynamous.	Epipetalous	Lobes of anthers not parallel.
Pistil. Carpels.	2	Syncarpous.	Superior.	

CHAPTER VIII.

EXAMINATION OF PLANTS WITH MONŒCIOUS AND DIŒCIOUS
FLOWERS—CUCUMBER, WILLOW.

52. **Cucumber.** You can hardly have failed to notice that only a small proportion of the blossoms on a Cucumber vine produce cucumbers. A great many wither away and are apparently of no use. An attentive inspection will show that some of the blossoms

have long fleshy protuberances beneath them, whilst
others are destitute of these
attachments. Select a flower
of each kind, and examine first
the one with the protuberance
(Fig. 61), which latter, from
its appearance, you will probably have rightly guessed to be
the ovary. The situation of the
ovary here, indeed, is the same
as in the Willow-herb. The

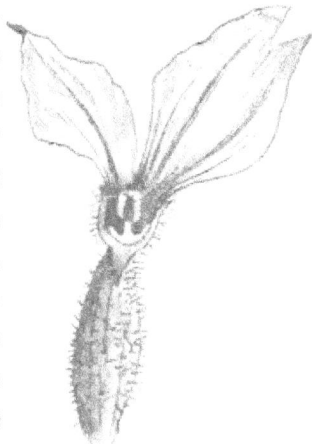

Fig. 61.

calyx-tube adheres to its surface, and is prolonged to
some little distance above it, expanding finally into five
teeth. The corolla is gamopetalous, and is adherent to
the calyx. Remove now the calyx and the adherent
corolla, and there is left in the centre of the flower a
short column, terminating in three stigmas, each two-
lobed.

There are no stamens.

53. Now examine the other blossom (Fig. 62).
Calyx and corolla have almost
exactly the same appearance as
before. Remove them, and you
have left three stamens growing on the calyx-tube, and
slightly united by their anthers
(syngenesious).

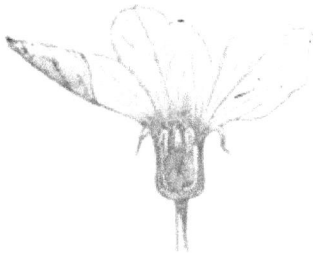

Fig. 62.

There is no pistil.

You see now why some blossoms produce cucumbers,
and others do not. Most of the blossoms have no
pistil, and are termed *staminate* or *sterile* flowers, whilst
the others are *pistillate* or *fertile.* Flowers in which

either stamens or pistils are wanting are also called *imperfect.* When staminate and pistillate flowers grow on
the same plant, as they do in the case of the Cucumber,
they are said to be *monœcious.*

54. In plants of this kind the pollen of one kind of
blossom is conveyed to the stigmas of the other kind,
chiefly by insects, which visit the flowers indiscrimin-
ately, in search of honey. The pollen dust clings to
their hairy legs and bodies, and is presently rubbed off
upon the stigma of some fertile flower.

55. In order to describe monœcious flowers, our
schedule will require a slight modification. As given
below, the symbol † stands for " staminate flower,"
and the symbol ‡ for " pistillate flower."

CUCUMBER.

ORGAN.	NO.	COHESION.	ADHESION.	REMARKS.
Calyx. *Sepals.*	5	Gamosepalous	Superior.	
Corolla. *Petals.*	5	Gamopetalous	**Perigynous.**	
† Stamens.	3	Syngenesious.	**Perigynous**	Two anthers are 2—celled, and one 1—celled.
† Pistil. *Carpels.*	0			
‡ Stamens.	0			
‡ Pistil. *Carpels.*	3	Syncarpous.	Inferior.	

56. **Willow.** The flowers of most kinds of Willow appear in spring or early summer, before the leaves. They grow from the axils in long close clusters called *catkins* or *aments*. Collect a few of these *from the same tree or shrub*. You will find them to be exactly alike. If the first one you examine is covered with yellow stamens (Fig. 63), all the rest will likewise consist of sta-mens, and you will search in vain

Fig. 63.

for any appearance of a pistil. If, on the other hand, one of your catkins is evidently destitute of stamens, and consists of oblong pis-tils (Fig. 64), then all the others will in like manner be found to be without stamens. Unlike our Cu-cumber plant, the stami nate and pistillate flowers of the Willow are borne

Fig. 64.

on *different* plants. These flowers are therefore said to be *dioecious*. As a general thing, staminate and pistil-late catkins will be found upon trees not far apart. Procure one of each kind, and examine first the stami-nate one. You will probably find the stamens in pairs. Follow any pair of filaments down to their insertion, and observe that they spring from the axil of a minute bract (Fig. 65). These bracts are the *scales* of the catkin. There is no appearance of either calyx or corolla, and the flowers are therefore said to be *achlamy-deous*, that is, without a covering. Now look

Fig. 65. at the fertile catkin. Each pistil will, like

the stamens, be found to spring from the axil
of a scale (Fig. 66). The stigma is two-lobed,
and on carefully opening the ovary you observe
that though there is but one cell, yet there are
two rows of seeds. We therefore infer that the
pistil consists of two carpels. The pistillate
flowers, like the staminate, are achlamydeous.
In diœcious plants, the process of fertilization
is assisted by insects, and also very largely by
the wind.

Fig. 66.

HEART-LEAVED WILLOW.

ORGAN	NO.	COHESION.	ADHESION.	REMARKS.
Calyx.	0			
Corolla.	0			
⊹ Stamens.	2	Diandrous.	0	
⊹ Pistil.	0			
: Stamens.	0			
: Pistil. Carpels.	2	Syncarpous.	0	

CHAPTER IX.

CHARACTERISTICS POSSESSED IN COMMON BY ALL THE PLANTS PREVIOUSLY EXAMINED. STRUCTURE OF THE SEED IN DICOTYLEDONS.

57. Before proceeding further in our examination of plants, we shall direct your attention to some characters of those already examined, which they all possess in common. The leaves of every one of them are *net-veined*. Some leaves, at least, of each of them have distinct petioles and blades. The parts of the flowers we found, as a general thing, to be in *fives*. In one or two instances they were in *fours*, that is, four sepals, four petals, and so on.

58. Now, in addition to these resemblances there are others which do not so immediately strike the eye, but which, nevertheless, are just as constant. One of these is to be found in the structure of the embryo. Take a cucumber or pumpkin seed, and having soaked it

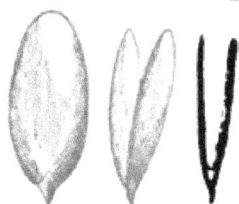

for some time in water, remove the outer coat. The body of the seed will then readily split in two, except where the parts are joined at one end. (Figs. 67, 68, 69). The thick lobes are called *cotyledons*, or *seed-*

Fig. 67. Fig. 68. Fig. 69.

leaves, and as there are two, the embryo is *dicotyledonous*. The pointed end, where the cotyledons are attached, and from which the root is developed, is called the *radicle*. Between the cotyledons, at the summit of the radicle, you will find a minute upward projection. This is a bud, which is known as the *plumule*. It developes into the stem.

59. If you treat a pea or a bean (Figs. 70, 71), in the same manner as the cucumber seed, you will find it to be

Fig. 70.

Fig. 71.

constructed on the same plan. The embryo of the bean is dicotyledonous also.
But you will observe that in these cases
the embryo occupies the whole of the interior of the seed. In describing the seed
of the Buttercup, it was pointed out that
the embryo occupies but a very small
space in the seed, the bulk of the lat-
ter consisting of *albumen*. Seeds like those of the Buttercup are therefore called *albuminous* seeds, while those of
the Bean and Pea are *exalbuminous*. But, notwithstand-
ing this difference in the structure of the seed, the *embryo* of
the Buttercup, when examined under a strong magnifier,
is found to be dicotyledonous like the others. In short,
the dicotyledonous embryo is a character common to all
the plants we have examined—common, as a rule, to
all plants possessing the other characters enumerated
above. From the general constancy of all these characters, plants possessing them are grouped together in
a vast **Class**, called Dicotyledonous plants, or, shortly,
Dicotyledons.

60. Besides the characters just mentioned, there is
still another one of great importance, which Dicotyledons possesses in common. It is *the manner of growth
of the stem*. In the Willow, and all our trees and shrubs
without exception, there is an outer layer of bark on the
stem, and the stem increases in thickness, year by year,
by forming a new layer just inside the bark and *outside
the old wood*. These stems are therefore called *exogenous*,
that is, *outside growers*.

Now, in all dicotyledonous plants, whether herbs,
shrubs or trees, the stem thickens in this manner, so
that **Dicotyledons** are also **Exogens**.

CHAPTER X.

EXAMINATION OF COMMON PLANTS CONTINUED. DOG'S-TOOTH
VIOLET, TRILLIUM, INDIAN TURNIP, CALLA, ORCHIS,
TIMOTHY.

61. **Dog's-tooth Violet.** This plant (Fig. 72)which
flowers in Spring, may be pretty easily recognised by

Fig. 72.

its peculiar blotched leaves. It may be found in rich

moist pasture lands and low copses. The name "Violet"
is somewhat unfortunate, because the plant is not in
any way related to the true Violets. To obtain a complete specimen requires some trouble, owing to the fact
that the root is commonly six inches or so below the
surface of the ground'; you must therefore insert a spade
or strong trowel sufficiently deep to avoid cutting or
breaking the tender stem. Having cleared away the
adhering earth, you will find that the roots proceed from
what appears to be the swollen end of the stem. This
swollen mass is coated on the outside with thin scales.
A section across the middle shows it to be more or less
solid, with the stem growing up through it from its
base. It is, in fact, not easy to say how much of this
stem-like growth is, in reality, stem, because it merges
gradually into the scape, which bears the flower, and the
petioles of the leaves, which sheathe the scape. The
swollen mass is called a *bulb*.

62. The leaves are two in number, gradually narrowing at the base into sheaths. If you hold one of them
up to the light, you will observe that the veins do not,
as in the leaves of the Dicotyledonous plants, form a
network, but run only in one direction, namely, from
end to end of the leaves. Such leaves are consequently
called *straight-veined*.

63. In the flower there is no appearance of a green
calyx. There are six yellow leaves, nearly alike, arranged in two sets, an outer and an inner, of three
each. In such cases, we shall speak of the colored
leaves collectively as the *perianth*. If the leaves are free
from each other, we shall speak of the perianth as *poly-
phyllous*, but if they cohere we shall describe it as *gamo-*

phyllous. Stripping off the leaves of the perianth we find six stamens, with long upright anthers which open along their outer edges. If the anthers be pulled off, the filaments will be found to terminate in long sharp points.

Fig. 73.

The pistil (Fig. 73) has its three parts,

Fig. 74.

ovary, style, and stigma, well marked. The stigma is evidently formed by the union of three into one. The ovary, when cut across, is seen to be three-celled (Fig. 74), and is therefore syncarpous.

DOG'S-TOOTH VIOLET.

ORGAN.	NO.	COHESION.	ADHESION.	REMARKS.
Perianth.		Polyphyllous.	Inferior.	
Leaves.	6			
Stamens.	6	Hexandrous.	Hypogynous.	Filaments terminating in sharp points.
Pistil.		Syncarpous.	Superior.	
Carpels.	3			

64. Trillium. This plant (Fig. 75) may be found in flower about the same time as the one just described. The perianth of Trillium consists of six pieces in two sets, but in this case the three outer leaves are green, like a common calyx. The stamens are six in number. There are three styles, curving outwards, the whole of the inner side of each being stigmatic.

The ovary (Fig. 76) is six-angled, and on being cut across is seen to be three-celled.

65. Comparing this flower with that of Dog's-tooth Violet, we find the two to exhibit a striking resemblance in structure. But in one respect the plants are strikingly unlike: *the leaves of the Trillium are net-veined* (Fig. 77), as in the Exogens. From this circumstance we learn that we cannot altogether rely on the veining of the leaves as a *constant* characteristic of plants whose parts are not in fives.

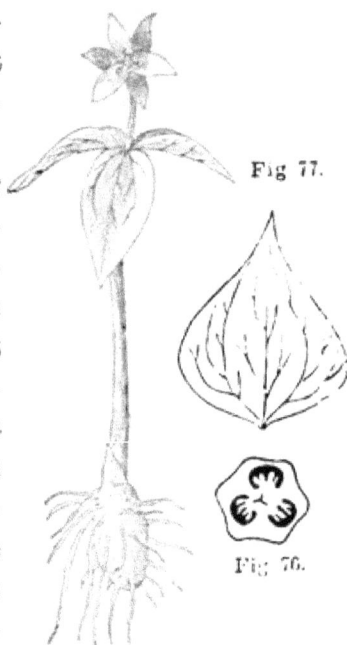

Fig 77.

Fig. 76.

Fig. 75.

TRILLIUM.

ORGAN.	NO.	COHESION.	ADHESION.	REMARKS.
Perianth.		Polyphyllous.	Inferior	Sepals persistent.
Sepals.	3			
Petals.	3			
Stamens.	6	Hexandrous.	Hypogynous.	
Pist'l.		Syncarpous.	Superior.	The inner face of each style stigmatic.
Carpels.	3			
Leaves net-veined.				

66. Indian Turnip. This plant may be easily met with in our woods in early summer. If you are not familiar with its appearance, the annexed cut (Fig. 78)

Fig. 78.

will help you to recognise it. Procure several specimens; these will probably at first seem to you to be alike in every respect, but out of a number, some are pretty sure to differ from the rest. Notice the bulb from which the stem springs. It differs from that of the Dog's-tooth Violet, and Lilies generally, in being a solid mass. It is called a *corm*. Between the pair of

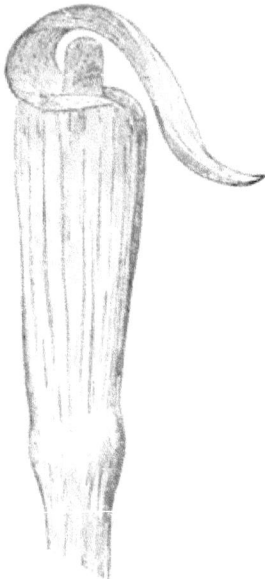
Fig. 79.

leaves you observe a curious striped sheath, having an arching, hood-like top, and enclosing an upright stalk, the top of which almost touches the hood (Fig. 79). Can this be a flower ? It is certainly the only thing about the plant which at all resembles a flower, and yet how different it is from any we have hitherto examined ! Carefully cut away the sheaths from all your specimens. Most, and perhaps all, of them will then present an appearance like that in Fig. 80. If none of them be like Fig. 81, it will be well to gather a few more plants. We shall suppose, however, that you have been fortunate in obtaining both kinds, and will proceed with our examination. Take first a specimen corresponding with Fig. 80. Around the base of the column are compactly arranged many spherical green bodies, each tipped with a little point. Separate one of these from the rest, and cut it across. It will be found to contain several ovules, and is, in fact, an ovary, the point at the top being a stigma. In the autumn, a great change will have taken place in the appearance of plants like the one we are now examining. The arched hood will have disappeared, as also the long naked top of the column, whilst the part below, upon which we are now engaged, will have

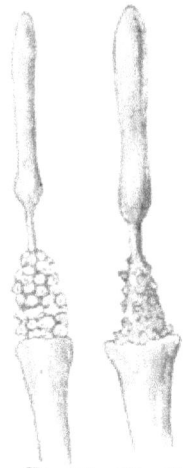
Fig. 80.   Fig. 81

vastly increased in size, and become a compact ball of red berries. There can be no doubt, then, that we have here a structure analogous to that found in the Cucumber and the Willow, the fertile, or pistillate, flowers being clustered together separately. But in the Cucumber all the flowers were observed to be furnished with calyx and corolla, and in the Willow catkins, though floral envelopes were absent, each pair of stamens and each pistil was subtended by a bract. In the present plant there are no floral envelopes, nor does each pistil arise from a separate bract.

67. But, you will now ask, what is this sheathing hood which we find wrapped about our column of pistils? There is no doubt that we must look upon it as a *bract*, because from its base the flower-cluster springs. So that, whilst the flowers of Indian-Turnip are, like those of Willow, imperfect and diœcious, the clusters differ in having but a single bract instead of a bract under each flower.

68. We must now examine one of the other specimens; and we shall have no difficulty in determining the nature of the bodies which, in this case, cover the base of the column. They are evidently stamens, and your magnifying-glass will show you that they consist mostly of anthers, the filaments being extremely short, and that some of the anthers are two-celled, and some four-celled, all discharging their pollen through little holes at the top of the cells.

69. The column upon which, in plants like Indian-Turnip, the flowers are crowded, is known as a *spadix*, and the surrounding bract as a *spathe*.

You will observe that the leaves of this plant are *net-veined*, as we found them in the Trillium.

INDIAN-TURNIP.

ORGAN.	NO.	COHESION.	ADHESION.
† Stamens.	1	Monandrous.	0
‡ Pistil.		Apocarpous.	0
Carpels.	1		

Flowers crowded on a spadix, and surrounded by a spathe.
Leaves net-veined.

70. **Marsh Calla.** This plant must be looked for in low marshy grounds, where it will be found in flower generally in the month of June. With the knowledge which you have of the structure of Indian-Turnip, you will hardly doubt that the Calla is closely related to it. You will easily recognize the spadix and the spathe (Fig. 82), though in the present instance the spadix

Fig. 82.

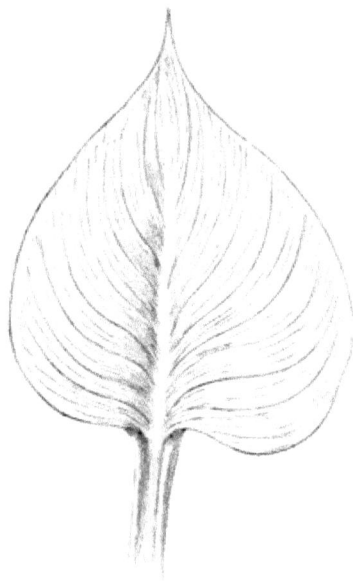

Fig. 83.

bears flowers to the ѡop, anɑ ѡne spathe is open instead
of enclosing the column. Observe, however, that the
veining of the leaf (Fig. 83) is different, that of Calla
being straight, like the Dog's-tooth Violet. There is
also a difference in the flowers. Those of Indian-Turnip were found to be diœcious, but the spadix, in the
present case, bears both stamens and pistils, and the
lower flowers, if not all, are *perfect ;* sometimes the upper ones consist of stamens
only. Fig. 84 shows one of the perfect
flowers much enlarged. The stamens, it
will be observed, have two-celled anthers,
opening lengthwise.

Fig. 84.

MARSH CALLA.

ORGAN.	NO.	COHESION.	ADHESION.
Perianth.		Wanting.	
Stamens	6	Hexandrous.	Hypogynous.
Pistil. Carpels.	1	Apocarpous.	Superior.

/ **71. Showy Orchis.** The flower of this plant
(Figs. 85, 86) is provided with floral envelopes, all coloured like a corolla. As in Dog's-tooth Violet, we shall
call them collectively the perianth, although they are
not all alike. One of them projects forward in front
of the flower, forming the *lip,* and bears underneath it a long hollow *spur,* which, like the spurs of
Columbine, is honey-bearing. The remaining five con-
verge together forming a kind of arch over the centre
of the flower. Each flower springs from the axil of ɑ

Fig. 85.

leaf-like bract, and is apparently raised on a pedicel.
What seems to be a pedicel, however, will, if cut across,
prove to be the ovary, which in this case is
inferior. Its situation is similar to the situation of the ovary in Willow-herb, and, as
in that flower, so in this the calyx-tube adheres to the whole surface of the ovary, and
the three outer divisions of the perianth are
simply upward extensions of this tube. Notice the peculiar *twist* in the ovary. The
effect of this twist is to turn the lip away

Fig. 86.

from the scape, and so give it the appearance of being the lower petal instead of the upper one, as it really is.

72. The structure of the stamens and pistils remains to be examined, and a glance at the flower shows you that we have here something totally different from the common arrangement of these organs. In the axis of the flower, immediately behind the opening into the spur, there is an upward projection known as the *column.* The face of this column is the stigma ; on each side of the stigma, and adhering to it, is an anther-cell. These cells, though separated by the column, constitute but a *single stamen.* The stamen, then, in this case is *united with the pistil,* a condition which is described as *gynan-drous.*

73. If you have a flower in which the anther-cells are bursting open, you will see that the pollen does not issue from them in its usual dust-like form, but if you use the point of your needle carefully you may remove the contents of each cell *in a mass.* These pollen masses are of the form shown in Fig. 87. The grains are kept together by a fine tissue or web, and the slender stalk, upon which each pollen mass is raised, is attached by its lower end to a sticky disk on the front of the stigma just above the mouth of the spur. Insects, in their efforts to reach the honey, bring their heads in contact with these disks, and when they fly away carry the pollen-masses with them, and deposit them on the stigma of the next flower visited. In fact, without the aid of insects it is difficult to see how flowers of this sort could be fertilized at all.

Fig. 87.

SHOWY ORCHIS.

ORGAN.	NO.	COHESION.	ADHESION.	REMARKS.
Perianth. *Leaves.*	6	Gamophyllous.	Superior.	
Stamens.	1	Monandrous.	Gynandrous.	Pollen-grains collected in masses.
Pistil. *Carpels.*	3	Syncarpous.	Inferior.	Ovary twisted.

74. **Timothy.** The top of a stalk of this well-known grass is cylindrical in shape, and upon examination will be found to consist of a vast number of similar pieces compactly arranged on very short pedicels about

the stalk as an axis. Carefully separate one of these pieces from the rest, and if the grass has not yet come into flower the piece will present the appearance shown in Fig. 88. In

Fig. 88. this Fig. the three points in the middle are the protruding ends of stamens. The piece which you have separated is, in fact, a flower enclosed in a pair of bracts, and all the other pieces which go to make up the top are flowers also, and, except perhaps a few at the very summit of the spike, precisely similar to this one in their structure.

75. Fig. 89 is designed to help you in dissecting a flower which has attained a greater degree of developement than the one shown in Fig. 88. Here the two bracts which enclose the flower have been drawn asunder. To these bracts the name *glumes* is applied. They are present in all

Fig 89.

plants of the Grass Family, and are often found enclosing several flowers instead of one as in Timothy. Inside the glumes will be found a second pair of minute chaff-like bracts, which are known as *palets* or *pales*. These enclose the flower proper.

76. The stamens are three in number, with the anthers fixed by the middle to the long slender filament. The anthers are therefore *versatile*. The styles are two in number, bearing long feathery stigmas. The ovary contains a single ovule, and when ripe forms a seed-like *grain*, technically known as a *caryopsis*.

TIMOTHY.

ORGAN.	NO.	COHESION.	ADHESION.
Glumes.	2		
Palets.	2		
Stamens.	3	Triandrous.	Hypogynous.
Pistil. Carpels.	1	Apocarpous.	Superior.

CHAPTER XI.

COMMON CHARACTERISTICS OF THE **PLANTS JUST EXAMINED.**
STRUCTURE **OF THE** SEED IN MONOCOTYLEDONS.

77. It is now to be pointed out that the six plants last examined, viz., Dog's-tooth Violet, Trillium, Indian Turnip, Calla, Orchis, and Timothy, though differing in various particulars, yet have some char-

ters common to all of them, just as the group ending
with Willow was found to be marked by characters
possessed by all its members. The flowers of Dicotyledons were found to have their parts, as a rule, in
fours or fives; those of our second group have them in
threes or *sixes,* never in fives.

78. Again, the leaves of these plants are straight-
veined, except in Trillium and Indian-Turnip, which
must be regarded as exceptional, and they do not as a
rule exhibit the division into petiole and blade which
was found to characterize the Exogens.

79. We shall now compare the structure of a grain
of Indian Corn with that of the Cucumber or Pumpkin
seed which we have already examined (page 45). It
will facilitate our task if we select a grain from an ear
which has been boiled. And first of all, let us observe
that the grain consists of something more than the
seed. The grain is very much like the achene of the
Buttercup, but differs in this respect, that the outer
covering of the former is completely united with the
seed-coat underneath it, whilst in the latter the true
seed easily separates from its covering. Remove the
coats of the grain, and what is left is a whitish starchy-
looking substance, having a yellowish body inserted in
a hollow (Fig. 90) in the middle of one side. This latter
body is the *embryo,* and may be easily removed. All
the rest is *albumen.* Fig. 91 is a front view of the

embryo, and Fig. 92 shows a vertical
section of the same. The greater part
of the embryo consists of a *single cotyledon.* The radicle is seen near the base.

Fig. 90. Fig. 91. Fig. 92. and the plumule above.

80. Comparing the result of our observations with

what we have already learned about the Cucumber seed, we find that whilst in the latter there are two cotyledons, in the present case there is but one, and this peculiarity is common to all the plants just examined, and to a vast number of others besides, which are consequently designated Monocotyledonous plants, or shortly Monocotyledons. The seeds of this great Class may differ as to the presence or absence of albumen, just as the seeds of Dicotyledons do, but in the number of their cotyledons they are all alike. The Orchids, however, are very peculiar from having no cotyledons at all.

81. In addition to the points just mentioned, viz : the number of floral leaves, the veining of the foliage leaves, the usual absence of distinct petioles, and the single cotyledon, which characterize our second great Class, there is still another, as constant as any of these, and that is, the mode of growth of the stem, which is quite at variance with that exhibited in Dicotyledonous plants. In the present group the increase in the thickness of the stem is accomplished not by the deposition of circle after circle of new wood outside the old, but by the production of new wood-fibres through the interior of the stem generally, and the consequent swelling of the stem as a whole. These stems are therefore said to be *endogenous*, and the plants composing the group are called Endogens, as well as Monocotyledons.

We shall explain more fully the structure of exogenous and endogenous stems, when we come to speak of the minute structure of plants in a subsequent chapter.

CHAPTER XII.

MORPHOLOGY OF ROOTS, STEMS, AND FOLIAGE-LEAVES.

82. From what has gone before, you should now be tolerably familiar with the names of the different organs of plants, and you have also had your attention directed to some modifications of those organs as they occur in different plants. In all these cases, the adjective terms, which botanists use to distinguish the variations in the form of the organs, have been placed before you, and if you have committed these carefully to memory, you will have laid a good foundation for the lessons which follow on **Morphology**, the name given to the study of the various forms assumed by the same organ in different plants, or in different parts of the same plant. In some instances, the terms employed, being derived from Latin and Greek, and specially devised for botanical purposes, may seem difficult to learn. We believe, however, that this difficulty will be found to be more apparent than real. You will be surprised at the ease with which the terms will occur to your mind if you learn them with the help of plants which are everywhere within your reach—if you be not satisfied with being mere book-botanists

With a good many terms you will find no difficulty whatever, since they will be found to have the same meaning in their botanical applications as they have in their everyday use.

83. The Root. This organ is called the descending axis of the plant, from its tendency to grow downward into the soil from the very commencement of its developement. Its chief use is to imbibe liquid nourishment, and transmit it to the stem. You will remember that in our examination of some common seeds, such as those of the Pumpkin and Bean (Figs. 67-71), we found at the junction of the cotyledons a small pointed projection called the *radicle.* Now, when such a seed is put into the ground, under favourable circumstances of warmth and moisture, it begins to grow, or *germinate,* and the radicle, which in reality is a minute stem, not only lengthens, in most cases, so as to push the cotyledons upwards, but developes a *root* from its lower extremity. All seeds, in short, when they germinate, produce roots from the extremity of the radicle, and roots so produced are called *primary* roots.

84. There are two well-marked ways in which a primary root may develope itself. It may, by the downward elongation of the radicle, assume the form of a distinct central axis, from the sides of which branches or fibres are given off, or root-fibres may spring in a cluster from the end of the radicle at the very commencement of growth. If the root grow in the first way, it will be a *tap-root* (Fig. 93), examples of which are furnished by the Carrot, the Mallow, and the Bean ; if in the second way, it will be a *fibrous* root, examples of which are furnished by the Buttercup (Fig. 1) and by the entire class of Monocotyledonous or Endogenous plants.

85. Tap-roots receive different names, ac-

Fig. 93.

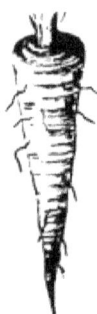

Fig. 94.

cording to the particular shape they happen to assume. Thus, the Carrot (Fig. 94) is *conical*, because from a broad top it tapers gradually and regularly to a point. The Radish, being somewhat thicker at the middle than at either end, is *spindle-shaped*. The Turnip, and roots of similar shape, are *napi_form* (*napus*, a turnip).

These fleshy tap-roots belong, as a rule, to biennial plants, and are designed as storehouses of food for the plant's use during its second year's growth. Occasionally fibrous roots also thicken in the same manner, as in the Peony, and then they are said to be *fascicled* or *clustered*. (Fig 95.)

86. But you must have observed that plants sometimes put forth roots in addition to those developed from the end of the radicle. The Verbena of of our gardens, for ex-

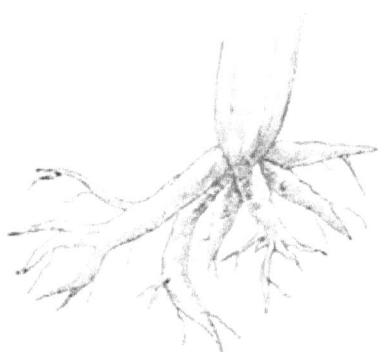

Fig. 95.

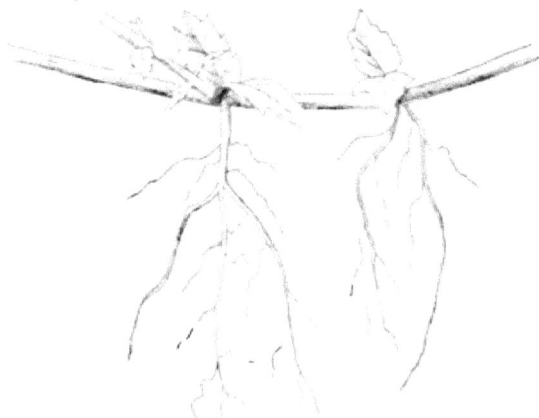

Fig. 96.

ample, will take root at every joint, if the stem be laid upon the ground (Fig. 96). The runners of the Strawberry take root at their extremities : and nothing is more familiar than that cuttings from various plants will make roots for themselves if put into proper soil, and supplied with warmth and moisture. All such roots are produced from some other part of the stem than the radicle, and are called *secondary* or *adventitious* roots. When such roots are developed from parts of the stem which are not in contact with the ground, they are *aerial*.

87. There are a few curious plants whose roots never reach the ground at all, and which depend altogether upon the air for food. These are called *epiphytes* There are others whose roots penetrate the stems and roots of other plants, and thus receive their nourish‑ ment as it were at second-hand. These are *parasitic* plants. The Dodder, Indian-Pipe, and Beech-drops, of Canadian woods, are well-known examples.

88. **The Stem.** As the root is developed from the lower end of the radicle of the embryo, so the stem is developed from the upper end, but with this important difference, that a *bud* always precedes the formation of the stem, or any part of it or its branches. Between the cotyledons of the Bean (Fig. 71), at the top of the radicle, we found a minute bud called the *plumule*. Out of this bud the first bit of stem is developed, and during the subsequent growth of the plant, wherever a branch is to be formed, or a main stem to be prolonged, there a bud will invariably be found. The branch buds are always in the axils of leaves, and so are called *axillary*. *Adventitious* buds, however, are sometimes produced in plants like the Willow, particularly if the

stem has been wounded. The bud from which the main stem is developed, or a branch continued, is of course at the end of the stem or branch, and so is *terminal.*

89. If you examine a few stems of plants at random, you will probably find some of them quite soft and easily compressible, while others will be firm, and will resist compression. The stem of a Beech or a Currant is an instance of the latter kind, and any weed will serve to illustrate the former. The Beech and the Currant have *woody* stems, while the weeds are *herbaceous.* Between the Beech and the Currant the chief difference is in size. The Beech is a *tree*, the Currant a *shrub.* But you are not to suppose that there is a hard and fast line between shrubs and trees, or between herbs and shrubs. A series of plants could be constructed, commencing with an unquestionable herb, and ending with an unquestionable tree, but embracing plants exhibiting such a gradual transition from herbs to shrubs, and from shrubs to trees, that you could not say at what precise point in the series the changes occurred.

90. The forms assumed by stems above ground are numerous, and they are described mostly by terms in common use. For instance, if a stem is weak, and trails along the ground, it is *trailing*, or *prostrate ;* and

Fig. 97.

if, as in the runners of the Strawberry, it takes root on the lower side, then it is *creeping.* Many weak stems raise

themselves by clinging to any support that may happen
to be within their reach. In some instances the stem
itself winds round the support, assuming a spiral form,
as in the Morning-Glory, the Hop, and the Bean, and
is therefore distinguished as *twining*. In other cases
the stem puts forth thread-like leafless branches called
tendrils (Fig. 97), which grasp the
support, as in the Virginia Creeper,
the Grape, and the Pea (Fig. 98), or
sometimes the leaf-stalks serve the
same purpose, as in the Clematis or
Virgin's Bower. In these cases the
stems are said to *climb*.

The stems of wheat and grasses
generally are known as *culms*. They
are jointed, and usually hollow except
at the joints.

Fig. 98.

91. Besides the stems which grow above ground, there

Fig. 99.

are varieties to be found below the surface. Pull up a

Potato plant, and examine the underground portion (Fig 99). It is not improbable that you will regard the whole as a mass of roots, but a very little trouble will undeceive you. Many of the fibres are unquestionably roots, but an inspection of those having potatoes at the ends of them will show you that they are quite different from those which have not. The former will be found to be furnished with little scales, answering to leaves, each with a minute bud in the axil; and the potatoes themselves exhibit buds of the same kind. The potato, in short, is only *the swollen end of an underground stem* Such swollen extremities are known as *tubers.* whilst the

Fig. 100.

underground stem is called a *rootstock,* or *rhizome,* and may always be distinguished from a true root by the presence of buds. The Solomon's Seal and Toothwort of Canadian woods, and the Canada Thistle, are common instances of plants producing these stems. Fig. 100 shows a rhizome.

92. Take now an Onion, and compare it with a Potato. You will not find any such outside appearances upon the former as are presented by the latter. The Onion is smooth, and has no buds upon its surface. From the under side there spring roots, and this circumstance will probably suggest that the Onion must be a stem of some sort. Cut the Onion through from top to bottom (Fig. 101). It will then be seen to be

made up of a number of coats. Strip off one or two, and observe that whilst they are somewhat fleshy where the onion is broadest they gradually become thinner towards the top. The long green tubes, which project from the top of the Onion during its growth, are, in fact, the prolongations of these coats. But the tubes are the leaves

Fig 101.

of the plant. The mass of our Onion, therefore, consists of the *fleshy bases of the leaves*. But you will observe that at the bottom there is a rather flat solid part upon which these coats or leaves are inserted, and which must consequently be a stem. Such a stem as this. with its fleshy leaves, is called a *bu''*. If the leaves form coats, as in the Onion, the bulb is *coated* or *tunicated;* if they do not, as in the lilies (Fig. 102), it is *scaly.*

-g. 102

93. Tubers and bulbs, then, consist chiefly of masses of nourishing matter ; but there is this difference, that, in the latter, the nourishment is contained in the fleshy leaves themselves, whilst, in the former, it forms a mass more or less distinct from the buds.

94. The thickened mass at the base of the stem of our Indian Turnip (Fig. 78) is more like a tuber than a bulb in its construction. It is called a *corm*, or solid bulb. The Crocus and Gladiolus of the gardens are other examples.

95. In the axils of the leaves of the Tiger Lily are produced small, black, rounded bodies, which, on examination, prove to be of bulbous structure. They are, in

fact, *bulblets*, and new plants may be grown from them.

96. Our Hawthorn is rendered formidable by the presence of stout *spines* (Fig. 103) along the stem and branches. These spines invariably proceed from the axils of leaves, and are, in fact, branches, whose growth has been arrested. They are appendages of the wood, and will remain attached to the stem,

Fig. 103.

even after the bark is stripped off. They must not be confounded with the *prickles* (Fig. 104) of the Rose and Brier, which belong strictly to the bark, and come off with it.

97. **Foliage-Leaves.** These organs are usually more or less flat, and of a green colour. In some plants, however, they are extremely thick and succulent; and in the case of parasites, such as Indian-Pipe and Beech-drops,

Fig. 104.

they are usually either white or brown, or of some colour other than green. The scaly leaves of underground stems are also, of course, destitute of colour.

98. As a general thing, leaves are extended horizontally from the stem or branch, and turn one side towards the sky and the other towards the ground. But some leaves are *vertical*, and in the case of the common Iris each leaf is doubled lengthwise at the base, and *sits astride* the next one within. Such leaves are accordingly called *equitant*.

99. As to their **arrangement** on the stem, leaves are *alternate* when only one arises from each node (Fig. 3). If two are formed at each node, they are sure to be

on opposite sides of the stem, and so are described as

opposite. Sometimes there are several leaves at the same node, in which case they are *whorled* or *verticillate* (Fig. 105).

100. Forms of Foliage-Leaves. Leaves present an almost endless variety in their forms, and accuracy in describing any given leaf depends a

Fig. 105.

good deal upon the ingenuity of the student in selecting and combining terms. The chief terms in use will be given here.

Compare a leaf of the Round-leaved Mallow with one of Red Clover (Figs. 106, 107). Each of them is fur-

Fig. 106.   Fig. 107.

nished with a long petiole and a pair of stipules. In the blades, however, there is a difference. The blade of the former consists of a *single piece*; that of the latter is in three *separate* pieces, each of which is called a *leaflet*, but all of which, taken collectively, constitute the blade

of the *leaf*.   The leaf of the Mallow is **simple**; that of
the Clover is compound.
Between the simple and
the compound form there
is every possible shade of
gradation.   In the Mallow
leaf the *lobes* are not very
clearly defined.   In the
Maple (Fig. 108) they are
well-marked. ·· In other
cases, again, the lobes are
so nearly separate, that

Fig. 108.

the leaves appear at first sight to be really compound.

101. You will remember that in our examinations of
dicotyledonous plants, we found the leaves to be invariably net-veined.   But, though they have this general character in common, they differ considerably in the
details of their veining, or **venation**, as it is called.
The two leaves employed as illustrations in the last
section will serve to illustrate our meaning here.   In
the Mallow, there are several ribs of about the same
size, radiating from the end of the petiole, something
like the spread-out fingers of a hand.   The veining in
this case is therefore described as *digitate*, or *radiate*, or
palmate.   The *leaflet* of the clover, on the other hand,
is divided exactly in the middle by a single rib (the
midrib), and from this the veins are given off on each
side, so that the veining, on the whole, presents the
appearance of a feather, and is therefore described as
pinnate (*penna*, a feather).

102. Both simple and compound leaves exhibit these
two modes of venation.   Of simple pinnately-veined

leaves, the Beech, Mullein, and Willow supply familiar instances. The Mallow, Maple, Grape, Currant, and Gooseberry have simple radiate-veined leaves. Sweet-Brier (Fig. 43), Mountain-Ash, and Rose have compound pinnate leaves, whilst those of Virginia-Creeper (Fig. 109), Horse-Chestnut, and Hemp are compound digitate.

Fig. 100.

As has already been pointed out, the leaves of Monocotyledonous plants are almost invariably straight veined.

103. In addition to the venation, the description of a simple leaf includes particulars concerning: (1) the general outline, (2) the edge or margin, (3) the point or apex, (4) the base.

104. Outline. As to outline, it will be convenient to consider first the forms assumed by leaves without lobes, and whose margins are therefore more or less continuous. Such leaves are of three sorts, viz: those in which both ends of the leaf are alike, those in which the apex is narrower than the base, and those in which the apex is broader than the base.

105. In the first of these three classes, it is evident that any variation in the outline will depend altogether on the relation between the length and the breadth of the leaf. When the leaf is extremely narrow in comparison with its length, as in the Pine, it is acicular or needle-shaped (Fig. 110). As the width increases, we pass through the forms known as linear, oblong, oval, and finally orbicular, in which the width and length are nearly, or quite equal (Fig. 111).

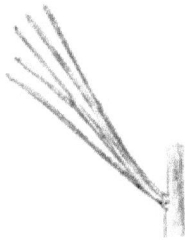

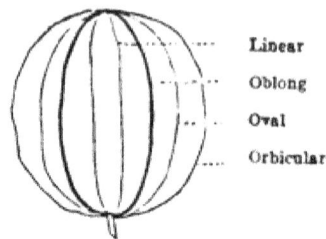

Fig. 110   Fig. 111.

106. In the second class the different forms arise from the varying width of the base of the leaf, and we thus have *subulate* or *awl-shaped* (Fig. 112), *lanceolar*. *ovate*, and *deltoid* leaves (Fig. 113).

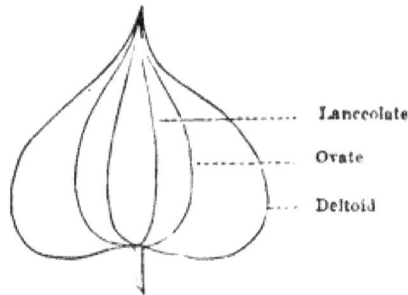

Fig. 112.   Fig. 113.

107. In the third class, as the apex expands, we have

Fig. 117.   Fig. 118.   Fig. 114.   Fig. 115.   Fig. 116.

the forms *spathulate* (Fig. **114**), *oblanceolate* (that is, the reverse of lanceolate) (Fig. **115**), and *obovate* (Fig. 116).

108. In leaves of the second kind we frequently find

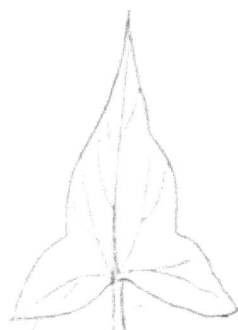

the base indented, and then the leaf is *cordate*, or *heart-shaped* (Fig. **117**). The reverse of this, that is, when the indentation is at the apex, is *obcordate* (Fig. **118**). The *hastate*, or spear-shaped (Fig. **119**), *sagittate*, or arrow-shaped (Fig. **120**), and *reniform*, or kidney-shaped (Fig. 121), forms are modifications of the second class,

Fig. 119.

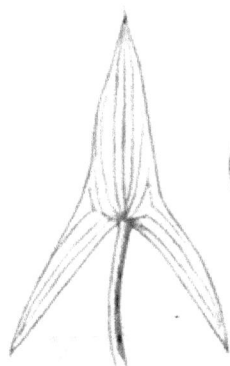

Fig. 120.   Fig. 121.   Fig. 122.

and will be readily understood from **the annexed** figures.

If the petiole is attached to any part of the under surface of the leaf, instead of to **the** edge, the leaf is *peltate* (shield-shaped) (Fig. 123).

109. Leaves which **are** lobed **are** usually described by stating whether they are palmately or pinnately veined,

Fig. 123.

and, if the former, the number of **lobes** is generally

given. If the leaves are very deeply cut, they are said
to be *palmatifid* or *pinnatifid* according to the veining
(Fig. 124). If the leaf is palmately lobed,
and the lobes at the base are themselves
lobed, the leaf is *pedate* (Fig. 125), because it looks something like a bird's
foot. If the lobes of a pinnatifid leaf
are themselves lobed, the leaf is *bipinnatifid*. If the leaf is cut up into fine
segments, as in Dicentra, it is said to be
multifid.

110. **Apex.** The principal forms of
the apex are the *mucronate* (Fig. 122),
when the leaf is tipped with a sharp

Fig. 124.

point, as though the midrib
were projecting beyond the
blade ; *cuspidate*, when the leaf
ends abruptly in a very short,
but distinctly tapering, point
(Fig. 126); *acute*, or sharp; and
obtuse, or blunt.

Fig. 125.

111. It may happen that the
apex does not end in a point of any kind. If
it looks as though the end had been cut off
square, it is *truncate*. If the end is slightly
notched, but not sufficiently so to warrant the
description obcordate, it is *emarginate*.

Fig. 126.

112. **Margin.** If the margin is not indented in any
way, it is said to be *entire*. If it has sharp teeth, *pointing in the direction of the apex*, it is *serrate*, and will be
coarsely or finely serrate, according to the size of the

Fig. 127.

teeth. Sometimes the edges of large teeth are themselves finely serrated, and in that case the leaf is *doubly serrate* (Fig. 127). If the teeth point *outwards*, that is, if the two edges of each tooth are of the same length, the leaf is *dentate*, but if the teeth, instead of being sharp, are *rounded*, the leaf is *crenate* (Fig. 128). The term *wavy* explains itself.

Fig. 128

113. **Base.** There are two or three peculiar modifications of the bases of simple sessile leaves which are of considerable importance in distinguishing plants. Sometimes a pair of lobes project backwards and cohere on the other side of the stem, so that the stem appears to pass through the leaf. This is the case in our common Bellwort, the leaves of which are accordingly described as *perfoliate* (Fig. 129). Sometimes two opposite sessile leaves grow together at the base, and clasp the stem, as in the upper leaves of Honeysuckle, in the Triosteum, and in one of our species of Eupatorium. Such leaves are said to be *connate* or *connate-perfoliate* (Fig. 130). In one of our Everlastings the margin of the leaf is continued on each side below the

Fig. 129.

Fig. 130.

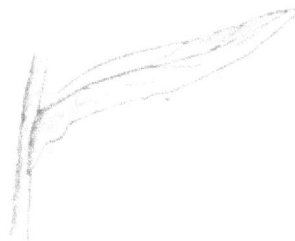

Fig. 131.

point of insertion, and the lobes grow fast to the sides

of the stem, giving rise to what is called the *decurrent* form (Fig. 131).

The terms by which simple leaves are described are applicable also to the leaflets of compound leaves, to the sepals and petals of flowers, and, in short, to any flat forms.

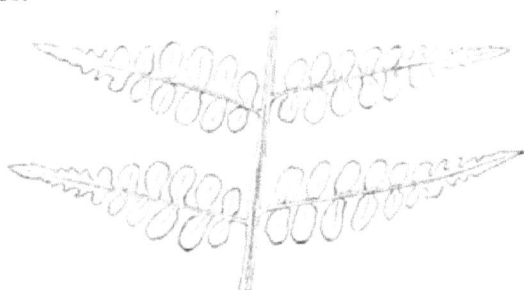

Fig. 132.

114. We have already explained that compound leaves are of two forms, *pinnate* and *palmate*. In the former, the leaflets are arranged on each side o the midrib. There may be a leaflet at the end, in which case the leaf is *odd-pinnate*, or the terminal leaflet may be wanting, and then the leaf is *abruptly pinnate*. In the Pea, the leaf is pinnate and terminates in a tendril (Fig. 98). Very frequently the primary divisions of a pinnate leaf are themselves pinnate, and the whole leaf is then *twice-pinnate* (Fig. 132). If the subdivision is continued through another stage, the leaf is *thrice-pinnate*, and so on. Sometimes, as in the leaves of the Tomato, very small leaflets are found between the larger ones, and this form is described as *interruptedly pinnate* (Fig. 132).

Fig. 133.

In the palmate or digitate forms, the leaflets spread out from the end of the petiole, and, in describing them, it is usual to mention the number of divisions. If there are three, the leaf is *tri-foliolate*; if there are five, it is *quinquefoliolate*.

115. In the examination of the Mallow, we found a couple of small leaf-like attachments on the petiole of each leaf, just at the junction with the stem. To these the name *stipules* was given. Leaves which have not these appendages are *exstipulate*.

116. Besides the characters of leaves mentioned above, there remain a few others to be noticed. With regard to their **surface**, leaves present every gradation from perfect smoothness, as in Wintergreen, to extreme roughness or woolliness, as in the Mullein. If hairs are entirely absent, the leaf is *glabrous*; if present, the degree of hairiness is described by an appropriate adverb; if the leaf is completely covered, it is *villous* or *villose*; and if the hairs are on the margin only, as in our Clintonia, it is *ciliate*. Some leaves, like those of Cabbage, have a kind of bloom on the surface, which may be rubbed off with the fingers; this condition is described as *glaucous*.

117. A few plants have **anomalous leaves**. Those of the Onion are *filiform*. The Pitcher Plant of our Northern swamps has very curious leaves (Fig. 134), apparently formed by the turning in and cohesion of the outer edges of an ordinary leaf, so as to form a tube, closed except at the top, and armed

on the inner surface with bristles pointing towards the base of the leaf.

118. Finally, as leaves present an **almost** infinite variety in their forms, it will often be necessary, in describing **them**, to **combine the** terms explained above. For instance, a leaf **may not** be exactly linear, **nor exactly** lance-shaped, but may approximate **to both forms.** In such **a case** the leaf is described as *lance-linear*, and so with other forms.

The following form of schedule **may** be used **with** advantage **in writing** out descriptions of leaves. **Two** leaves—one of Maple and one **of** Sweet-Brier—are described **by way** of illustration. If a leaf is compound, the particulars as to outline, margin, apex, **base, and** surface will have reference to the leaflets.

LEAF SCHEDULE.

LEAF OF	MAPLE.	SWEET-BRIER.
1. Position.	Cauline.	Cauline.
2. Arrangement.	Opposite.	Alternate.
3. Insertion.	Petiolate.	Petiolate.
4. Stipulation.	Exstipulate.	Stipulate.
5. Division.	Simple.	Odd pinnate, 7 leaflets
6. Venation.	Palmate.	
7. Outline.		Roundish or oval.
8. Margin.	Deeply lobed.	Doubly serrate.

9. Apex.	Pointed.	Acute.
10. Base.	Cordate.	Hardly indented.
11. Surface.	Glabrous above; whitish beneath.	Downy above; covered with glands beneath.

CHAPTER XIII.

MORPHOLOGY OF FLOWER-LEAVES. THE CALYX. THE CO-
ROLLA. THE STAMENS. THE PISTIL. THE FRUIT.
THE SEED. GERMINATION.

119. From an examination of the various forms presented by foliage leaves, we proceed now to those of
the floral ones, and we shall first consider the chief
modifications in the *arrangement of flowers as a whole*,
to which the term **inflorescence** is applied.

120. It is found that inflorescence proceeds upon two
well-defined plans. To understand these, let us recur
to our specimens of Shepherd's-Purse and Buttercup.
You will remember that, in the former, the peduncle
continues to lengthen as long as the summer lasts, and
new flowers continue to be produced at the upper
end. Observe, however, that every one of the flowers is
produced *in the axil of a bract*, that as the stem lengthens
new bracts appear, and that *there is no flower on the end
of the stem*. You will easily understand then, that
the production of flowers in such a plant is only
limited by the close of the season or by the exhaus-
tion of the plant. Such inflorescence is therefore
called indefinite, or indeterminate, or axillary.
It is sometimes also called *centripetal*, because if the
flowers happen to be in a close cluster, as are the upper

ones in Shepherd's-Purse, the order of developement is from the outside *towards the centre*.

121. If you now look at your Buttercup, you will be at once struck with the difference of plan exhibited. Tho main axis or stem has *a flower on the end of it*, and its further growth is therefore checked. And so in like manner, from the top downwards, the growth of tho branches is checked by the production of flowers at their extremities. The mode of inflorescence here displayed is **definite**, or **determinate**, or **terminal**. It is also called *centrifugal*, because the developement of the flowers is the reverse of that exhibited in the first mode. The upper, or, in the case of close clusters, the *central* flowers open first. In either mode, if there is but one flower in each axil, or but one flower at the end of each branch, the flowers are said to be *solitary*.

122. Of **indeterminate inflorescence** there are several varieties. In Shepherd's-Purse we have an instance of the *raceme*, which may be described as a cluster in which each flower springs from an axil, and is supported on a pedicel of its own. If the pedicels are absent, and the flowers consequently sessile in the axils, the cluster becomes a *spike*, of which the common Plantain and the Mullein furnish good examples. The *catkins* of the Willow (Figs. 63, 64) and Birch, and the *spadix* of the Indian Turnip (Figs. 80, 81) are also spikes, the former having scaly bracts and the latter a fleshy axis. If you suppose the internodes of a spike to be suppressed, so that the flowers are densely crowded, you will have a *head*, of which Clover and Button-bush supply instances. If the lower pedicels of a raceme are considerably longer than the upper ones, so that all the blossoms are nearly on the same

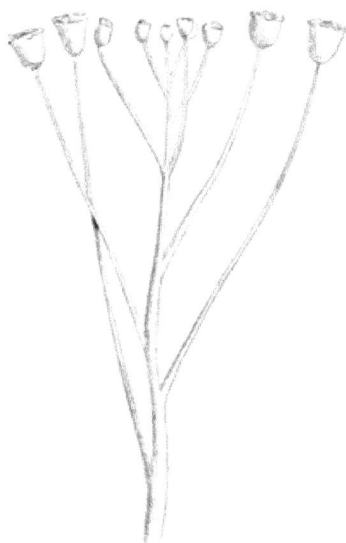

Fig. 135.　　　　　　　　Fig. 136.

level, the cluster is a *corymb* (Fig. 135). If the flowers in a head were elevated on separate pedicels of the same length, radiating like the ribs of an umbrella, we should have an *umbel*, of which the flowers of Geranium and Parsnip (Fig. 49) are examples. A raceme will be *compound* (Fig. 136) if, instead of a solitary flower, there is a *raceme in each axil*, and a similar remark will apply in the case of the spike, the corymb, and the umbel.

123. The inflorescence of most Grasses is what is called a *panicle*. This is a compound form, and is usually a kind of raceme having its **primary** divisions branched in some irregular manner. If the panicle is

compact, as in the Grape and Lilac, it is what is called a *thyrse*.

124. Of **determinate inflorescence** the chief modification is the *cyme*. This is a rather flat-topped cluster, having something the appearance of a compound corymb, but easily distinguished by this peculiarity, that the *central blossom opens first*, then those at the ends of the first set of branches of the cluster, then those on the secondary branches, and so on until the outer buds are reached. The Elder, Dogwood, and St. John's Wort furnish good examples of the cymose

Fig. 137.

structure. Fig. 137 shows a loose open cyme.

125. It has already been pointed out that cauline leaves tend to diminish in size towards the upper part of the stem, where the flowers are found. Such reduced leaves, containing flowers in their axils, are called *bracts*. In the case of compound flower-clusters, this term is limited to the leaves on the peduncle, or main stem, the term *bractlet* being then applied to those occurring on the pedicels or subordinate stems. In the

case of the *umbel* and the *head*, it generally happens that a *circle* of bracts surrounds the base of the cluster. They are then called, collectively, an *involucre*, and in the case of compound clusters a circle of bractlets is called an *involucel*. Bracts are often so minute as to be reduced to mere *scales*. From our definition, it will be evident that the *spathe* surrounding the spadix in Indian Turnip is merely a bract.

126. It has already been stated that the parts of the flower, equally with the foliage-leaves, must be regarded as modifications of the same structure, and some proofs of this similarity of structure were given. We shall now proceed to consider in detail the variations in form assumed by these organs.

127. **The Calyx.** As you are now well aware, this term is applied to the outer circle of floral leaves. These are usually green, but not necessarily so; in some Exogens, and in nearly all Endogens, they are of some other colour. Each division of a calyx is called a *sepal*, and if the sepals are entirely distinct from each other, the calyx is *polysepalous*; if they are united in any degree, it is *gamosepalous*. A calyx is *regular* or *irregular*, according as the sepals are of the same or different shape and size.

128. In a gamosepalous calyx, if the sepals are not united to the very top, the free portions are known as *calyx-teeth*, or, taken collectively, as the *limb* of the calyx. The united portion, especially if long, as in Willow-herb, is called the *calyx-tube*, and the entrance to the tube its *throat*. In many plants, particularly those of the Composite Family, the limb of the calyx consists merely of a circle of bristles or soft hairs.

and is then described as *pappose*. In other cases the limb is quite inconspicuous, and so is said to be *obsolete*. A calyx which remains after the corolla has disappeared, as in Mallow (Fig. 31), is *persistent*. If it disappears when the flower opens, as in our Bloodroot, it is *caducous*, and if it falls away with the corolla, it is *deciduous*.

We must repeat here, that when calyx and corolla are not both present, the circle which is present is considered to be the calyx, whether green or not.

129. **The Corolla.** The calyx and corolla, taken together, are called the *floral envelopes*. When both envelopes are present, the corolla is the inner cne; it is usually, though not invariably, of some other colour than green. Each division of a corolla is called a *petal*, and the corolla is *polypetalous* when the petals are completely disconnected; but *gamopetalous* if they are united in any degree, however slight. The terms *regular* and *irregular*, applied to the calyx, are applicable also to the corolla, and the terms used in the description of leaves are applicable to petals. If, however, a petal is narrowed into a long and slender portion towards the base, that portion is known as the *claw*, whilst the broader upper part is called the *limb* (Fig. 138). The leaf-terms are then applicable to the limb.

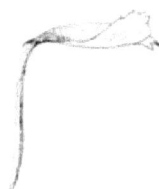

Fig. 138.

130. Gamopetalous corollas assume various forms, most of which are described by terms easily understood. The forms assumed depend almost entirely on the shape of the petals which, when united, make up the corolla. If these, taken separately, are linear, and are

united to the top, or nearly so, the corolla will be tubular (Fig. 139.) If the petals are wedge-shaped, they will by their union produce a *funnel-shaped* corolla. (Fig. 140.) In the *campanulate* or *bell-shaped* form, the enlargement from base to summit is more gradual. If the petals are narrowed abruptly into long claws, the union of the claws into a tube and the spreading of the limb at right

Fig 139

angles to the tube will produce the *salver-shaped* form, as in Phlox (Fig. 141). The *rotate* corolla differs from this in having a *very short* tube. The corolla of the Potato is rotate.

131. The most important *irregular* gamopetalous corollas are the *ligulate*, which has been fully described in the examination of the Dandelion, and the *labiate*, of which we found an example in

Fig. 140

Catnip (Fig. 59). The corolla of Turtle-head (Fig. 142) is another example. When a labiate corolla presents a wide opening between the upper and lower lips, it is said to be *ringent*. if the opening is closed by an

Fig 141.

Fig. 142.

Fig. 143.

upward projection of the lower lip, as in Toadflax (Fig. 143), it is said to be *personate*, and the projection in

this case is known as the *palate*. A good many corol
las such as those of Toadflax, Dicentra, Snapdragon,
Columbine, and Violet, have protuberances or *spurs* at
the base. In Violet one petal only is spurred; in
Columbine the whole five are so.

132. **The Stamens.** As calyx and corolla are
called collectively the floral envelopes, so stamens and
pistil are spoken of collectively as the *essential organs* of
the flower. The circle of stamens alone is sometimes
called the *andræcium*. A complete stamen consists of a
slender stalk known as the *filament*, and a small sac
called the *anther*. The filament, however, is not un-
commonly absent, in which case the anther is *sessile*.
As a general thing, the anther consists of two oblong
cells with a sort of rib between them called the *connective*, and that side of the anther which presents a distinctly *grooved* appearance is the *face*, the opposite side
being the *back*. The filament is invariably attached to
the connective, and may adhere through the entire
length of the latter, in which case the anther is *adnate*

Fig. 144.   Fig. 145.    Fig. 146.

(Fig. 144), or the base of the connective may rest on
the end of the filament, a condition described as *innate*
(Fig. 145), or the extremity of the filament may be
attached to the middle of the back of the connective, so
that the anther swings about; it is then said to be
versatile (Fig. 146). In all these cases, if the face of

the anther is turned towards the centre of the flower, it is said to be *introrse*; if turned outwards, *extrorse*.

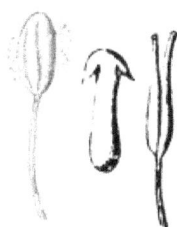

133. The cells of anthers commonly open along their outer edges to discharge their pollen (Fig. 147). In most of the Heaths, however, the pollen is discharged through a minute aperture at the top of each cell (Fig. 148), and in our Blue Cohosh each cell is provided with a lid or valve near the top, which opens on a kind of hinge (Fig. 149).

134. Stamens may be either entirely distinct from each other, in which case they are described as *diandrous*, *pentandrous*, *octandrous*, &c., according to their number (or, if more than twenty, as indefinite), or they may be united in various ways. If their anthers are united in a circle, while the filaments are separate (Fig. 81), they are said to be *syngenesious*, but if the filaments unite to form a tube, while the anthers remain distinct, they are said to be *monadelphous* (Fig. 82); if they are in two groups they are *diadelphous* (Fig. 87); if in three, *triadelphous*; if in more than three, *polyadelphous*.

135. As to insertion, when stamens are inserted on the receptacle, they are *hypogynous;* when borne on the calyx, *perigynous* ; when borne on the ovary, *epigynous;* and if inserted on the corolla, *epipetalous*. They may, however, be borne even on the style, as in Orchis, and then they are described as *gynandrous*.

136. If the stamens are four in number, and in two pairs of different lengths, they are said to be *didynamous* (Fig. 58); if six in number, four long and two short, they are *tetradynamous* (Fig. 59), and, finally,

if the stamens are hidden in the tube of a gamopetalous corolla, they are said to be *included*, but if they protrude beyond the tube they are *exserted* (Fig. 139).

137. **The Pistil.** This is the name given to the central organ of the flower. It is sometimes also called the *gynœcium*. As in the case of the stamens, the structure of the pistil must be regarded as a modification of the structure of leaves generally. The pistil may be formed by the folding of a single carpellary leaf as in the Bean (Fig. 159), in which case it is *simple*; or it may consist of a number of carpels, either entirely separate from each other, or united together in various ways, in which case it is *compound*. If the carpels are entirely distinct, as in Buttercup, the pistil is *apocarpous*; if they are united in any degree, it is *syncarpous*.

138. In our examination of the Marsh Marigold (Figs. 24, 25) we found an apocarpous pistil of several carpels. We found also that each carpel contained a number of seeds, and that, in every case, the seeds were attached to that edge of the carpel *which was turned towards the centre of the flower*, and that, as the carpels ripened, they invariably split open along that edge, but not along the other, so that the carpel when opened out presented the appearance of a leaf with seeds *attached to the margins*. The inner edge of a simple carpel, to which the seeds are thus attached, is called the *ventral suture*, the opposite edge, corresponding to the mid-rib of a leaf, being the *dorsal suture*.

139. If we suppose a number of simple carpels to approach each other, and unite in the centre of a flower, it is evident that the pistil so formed would contain as many cells as there were carpels, the cells being separ-

ated from each other by a *double wall*, and that the
seeds would be found arranged about the centre or axis
of the pistil ; and this is the actual state of things in
the Tulip, whose pistil is formed by the union of three
carpels. When the pistil ripens, the double walls separating the cells split asunder. To these separating
walls the name *dissepiment* or *partition* is given.

140. But it often happens that though several carpels unite to form a compound pistil, there is but one
cell in the ovary. This is because the separate carp
lary leaves have not been folded before uniting, :
have been joined edge to edge, or
rather with their edges slightly
turned inwards. In these cases the
seeds cannot, of course, be in the
centre of the ovary, but will be found

Fig. 151.　　Fig. 150.

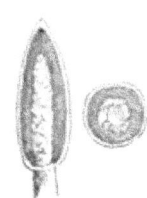

on *the walls*, at the junction of the carpels
(Figs. 150, 151). In some plants the ovary
is one-celled, and the seeds are arranged
round a column which rises from the bottom
of the cell (Figs. 152, 153). This case is

Figs 152, 153. explained by the early obliteration of the
partitions, which must at first have met in the centre of
the cell.

141. In all cases the line or projection to which the
seeds are attached is called the *placenta*, and the term
placentation has reference to the manner in which
the placentas are arranged. In the simple pistil the
placentation is *marginal* or *sutural*. In the **syncarpous**
pistil, if the dissepiments meet in the centre of the
ovary, thus dividing it into separate cells, the placentation is *central* or *axile ;* if the ovary is one-celled and
bears the seeds on its walls, the placentation is *parieto'.*

and if the seeds are attached to a central column, it is *free central.*

142. Besides the union of the ovaries there may also be a union of the styles, and even of the stigmas.

143. A very exceptional pistil is found in plants of the Pine Family. Here the ovules, instead of being enclosed in an ovary, are usually simply attached to the inner surface of an open carpellary

Fig. 154.

leaf or scale, the scales forming what is known as a *cone* (Figs. 154, 155, 156). The plants of this family are hence called Figs. 155, 156. *gymnospermous*, or naked-seeded.

144. **The Fruit.** In coming to the consideration of the Fruit, you must for the present lay aside any popular ideas you may have acquired as to the meaning of this term. You will find that, in a strict botanical sense, many things are fruits which, in the language of common life, are not so designated. For instance, we hardly speak of a pumpkin or a cucumber as fruit, and yet they are clearly so, according to the botanist's definition of that term. A fruit may be defined to be *the ripened pistil together with any other organ, such as the calyx or receptacle, which may be adherent to it.* This definition will perhaps be more clearly understood after a few specimens have been attentively examined.

145. For an example of the simplest kind of fruit let us revert to our Buttercup. As the carpels ripen, the style and stigma are reduced to a mere point. On cutting open one of these carpels when fully ripe, we find it contains a single seed, not quite filling the cavity, but attached at one point to the wall of the latter. What you have to guard against, in this

instance, is the mistake of considering the entire
carpel to be merely a seed. It is a seed enveloped in an outer covering which we called the ovary
in the early stages of the flower, but which, now
that it is ripe, we shall call the *pericarp*. This pericarp,
with the seed which it contains, is the fruit. The principal difference between the fruit of Marsh-Marigold
and that of Buttercup is, that, in the former, the pericarp envelopes several seeds, and, when ripe, *splits open
down one side*. The fruit of Buttercup does not thus
split open. In the Pea, again, the pericarp encloses
several seeds, but splits open along *both* margins. The
fruits just mentioned all result from the ripening of
apocarpous pistils, and they are consequently spoken of
as *apocarpous fruits*.

146. In Willow-herb, you will recollect that the
calyx-tube adheres to the whole surface of the ovary.
The fruit in this case, then, must include the calyx.
When the ovary ripens, it splits longitudinally into four
pieces (Fig. 41), and, as the pistil was *syncarpous*, so
also is the fruit.

147. In the Peach, Plum, Cherry, and *stone-fruits* or
drupes generally, the seed is enclosed in a hard
shell called a *putamen*. Outside the putamen is
a thick layer of pulp, and outside this, enclosing
the whole, is a skin-like covering. In these fruits
all outside the seeds is the pericarp. In one
respect these stone-fruits resemble the fruit of the
Buttercup: they do not split open in order to discharge
their seeds. All fruits having this peculiarity are said
to be **indehiscent**, whilst those in which the pericarp
opens, or separates into pieces (called *valves*), are de-
hiscent.

148. In the Apple (Fig. 48) and Pear, the seeds are contained in five cells in the middle of the fruit, and these cells are surrounded by a firm fleshy mass which is an enlargement of the calyx. In fact, the remains of the five calyx-teeth may be readily detected at the end of the apple opposite the stem. As in Willow-herb, the calyx is adherent to the ovary, and therefore calyx and ovary together constitute the pericarp. These *fleshy-fruits*, or *pomes*, as they are sometimes called, are of course *indehiscent*.

149. In the Currant, as in the Apple, you will find the remains of a calyx at the top, so that this fruit, too, is *inferior*, but the seeds, instead of being separated from the mass of the fruit by tough cartilaginous cell-walls, as in the Apple, lie imbedded in the soft juicy pulp. Such a fruit as this is a *berry*. The Gooseberry and the Grape are other examples. The Pumpkin and other *gourds* are similar in structure to the berry, but besides the soft inner pulp they have also a firm outer layer and a hard rind. The name *pepo* is generally given to fruits of this sort.

Fig. 157.

150. A Raspberry or Blackberry (Fig. 157) proves, on examination, to be made up of a large number of juicy little drupes, aggregated upon a central axis. It cannot, therefore, be a true berry, but may be called an *aggregated* fruit.

151. A Strawberry (Fig. 158) is a fruit consisting chiefly of a mass of pulp, having its surface dotted over with little carpels (achenes) similar to those of the Buttercup. The flesh of the Strawberry

Fig. 158.

is simply an enlarged *receptacle ;* so that this fruit, also, is not a true berry.

152. The fruit of Sweet-Brier (Fig. 45) consists of a red fleshy calyx, lined with a hollow receptacle which bears a number of achenes. This fruit is therefore analogous to that of the Strawberry. In the latter the achenes are on the outer surface of a *raised* receptacle, while, in the former, they are on the inner surface of a *hollow* receptacle.

153. The Cone of the Pine (Fig. 154) is a fruit which differs in an important respect from all those yet mentioned, inasmuch as it is the product, not of a single flower, but of as many flowers as there are scales. It may therefore be called a *collective* or *multiple* fruit. The Pine-Apple is another instance of the same thing.

154. Of dehiscent fruits there are some varieties which receive special names. The fruit of the Pea, or Bean (Fig. 159), whose pericarp splits open along *both* margins, is called a *legume ;* that of Marsh-Marigold (Fig. 25), which opens down

Fig. 159.

one side only, is a *follicle.* Both of these are apocarpous.

155. Any syncarpous fruit, having a dry *dehiscent* pericarp, is called a *capsule.* A long and slender capsule, having two cells separated by a membranous partition bearing the seed, and from which, when ripe, the valves fall away on each side, is called a *silique*

(Fig. 160). If, as in Shepherd's Purse (Fig. 29), the capsule is short and broad, it is called a *silicle*. If the capsule opens *horizontally*, so that the top comes off like a lid, as in Purslane (Fig. 161), it is a *pyxis*.

Fig. 161.

156. Any dry, one-seeded, *indehiscent* fruit is called an *achene*, of which the fruit of Buttercup (Fig. 14) is an example. In Wheat the fruit differs from that of Butter-cup in having a closely fitting and adherent pericarp. Such a fruit is called a *caryopsis* or *grain*. A *nut* is usually syncarpous, with a hard, dry peri-carp. A *winged* fruit, such as that of the Maple (Fig. 162), is called a *samara* or *key*.

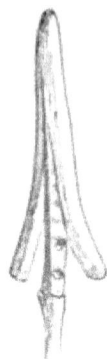

Fig. 160.

Fig. 162.

157. **The Seed**. The seed has already been de-scribed as the *fertilized ovule*. It consists of a *nucleus*, enveloped, as a rule, in two coats. The outer one, which is the most important, is known as the *testa*. Occasionally an additional outer coat, called an *aril*, is found. In the Euonymus of Canadian woods, the aril is particularly prominent in autumn, owing to its bright scarlet colour. The stalk, by which the seed is attached to the placenta, is the *funiculus*, and the scar, formed on the testa where it separates from the seed-stalk, is called the *hilum*. In the Pea and the Bean this scar is very distinct.

158. **Germination of the Seed**. When a seed is lightly covered with earth, and supplied with warmth and moisture, it soon begins to swell and soften, owing to the absorption of water, and presently bursts its

coats, either to such a degree as to liberate the cotyle·
dons completely, or so as to permit the escape of the
radicle and the plumule. The former immediately
takes a downward direction, developing a root from its
lower end, and either elongates through its whole
length, in which case the cotyledons are pushed above
the surface, as in the Bean, or remains stationary, in
which event the cotyledons remain altogether under
ground, as in the Pea and in Indian Corn.

Before the root is developed, and the little plantlet
is thereby enabled to imbibe food from the soil, it has to
depend for its growth upon a store of nourishment
supplied by the parent plant before the seed was cast
adrift. The relation of this nourishment to the embryo
is different in different seeds. In the Bean and the
Pumpkin, for example, it is contained in the cotyledons
of the embryo itself. But in Indian Corn, as we have
already seen, it constitutes the bulk of the seed, the
embryo merely occupying a hollow in one side of it. In
such cases as the latter, it will be remembered that the
term *albumen* is applied to the nourishing matter, as
distinguished from the embryo.

159. As to the number of cotyledons, it may be repeated that, as a rule, seeds are either dicotyledonous
or monocotyledonous. Some plants of the Pine Fam·
ily, however, exhibit a modification of the dicotyledonous structure, having several cotyledons, and being
consequently distinguished as *polycotyledonous*.

CHAPTER XV.

ON THE MINUTE STRUCTURE OF PLANTS—EXOGENOUS AND
ENDOGENOUS STEMS—FOOD OF PLANTS.

160. Up to this point we have been engaged in
observing such particulars of structure in plants as are
manifest to the naked eye. It is now time to enquire a
little more closely, and find out what we can about the
elementary structure of the different organs. We have
all observed how tender and delicate is a little plantlet
of any kind just sprouting from the seed; but as time
elapses, and the plant developes itself and acquires
strength, its substance will, as we know, assume a
texture varying with the nature of the plant, either
becoming hard and firm and woody, if it is to be a tree
or a shrub, or continuing to be soft and compressible
as long as it lives, if it is to be an herb. Then, as a
rule, the leaves of plants are of quite a different consistency from the stems, and the ribs and veins and
petioles of foliage leaves are of a firmer texture than
the remaining part of them. In all plants, also, the
newest portions, both of stem and root, are extremely
soft compared with the older parts. It will be our
object in this chapter to ascertain, as far as we can, the
reason of such differences as these; and to accomplish
this, we shall have to call in the aid of a microscope of
much higher power than that which has hitherto
served our purpose.

161. If a small bit, taken from a soft stem, be boiled
for a while so as to reduce it to a pulp, and a little of
this pulp be examined under the microscope, it will be
found to be entirely composed of more or less rounded

or oval bodies, which are either loosely thrown together
(Fig. 163), or are pressed into a more or less compact

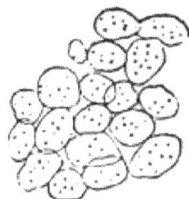  

Fig. 163.   Fig. 164.   Fig. 164 (a).

mass. In the latter case, owing to mutual pressure,
they assume a somewhat angular form. These bodies
are called *cells*. They are hollow, and their walls are
usually thin and transparent. The entire fabric of
every plant, without any exception whatever, is made
up of cells ; but as we proceed in our investigation, we
shall find that these cells are not all precisely alike,
that as they become older they tend, as a rule, to
thicken their walls and undergo changes in form, which,
to a great extent, determine the texture of the plant's
substance.

162. A fabric made up of cells is called a *tissue*. A
collection of such cells as we found constituting our
pulp, and as we should find constituting the mass of
all the soft and new parts of plants, as well as of some
hard parts, is called *cellular tissue*. The cells composing cellular tissue vary a great deal in size in different
plants, being, as a rule, largest in aquatics, in which
they may sometimes be observed with the naked eye.
Ordinarily, however, they are so minute that millions
of them find room in a cubic inch of tissue.

163. When young, the walls of the cells are quite
unbroken. Each cell is lined with an extremely thin
membrane, and a portion of its cavity is occupied by a

soft body called the *nucleus*. The space between the nucleus and the lining of the cell is filled with a thickish liquid called *protoplasm*, and the microscope reveals to us the fact that, as long as the cells are *living* cells, a circulation or current is constantly kept up in the protoplasm of each. To this curious movement the term *cyclosis* has been applied. As the cells become older, the nature of their contents is altered by the introduction of watery sap, in which other substances are found, notably starch, sugar, chlorophyll (to which leaves owe their green colour), and crystals (raphides) of various salts of lime. The substance of which the cell-wall is composed is called *cellulose*, and is a chemical compound of carbon, hydrogen, and oxygen. In the protoplasm nitrogen is found in addition to the three elements just mentioned.

164. *The growth of a plant consists in the multiplication of its cells.* Every plant begins its existence with a single cell, and by the repeated division of this, and the growth of the successive sections, the whole fabric of the plant, whether herb, shrub, or tree, is built up. The division of a cell is accomplished by the formation of a partition across the middle of it, the nucleus having previously separated into two pieces. The partition is formed out of the *lining* of the cell. Each half of the cell then enlarges, and, when its full size is attained, divides again, and so on, as long as the cells are *living* cells.

165. But in order to increase their size, *food* of some kind is essential. Growing plants supply this to their cells mainly in the form of sap, which is taken in by the root-fibres, and made suitable, or *elaborated*, or *assimilated*, by chemical action in the plant itself. By

a very curious process, the liquids absorbed by the root
pass from cell to cell, though each is quite enclosed,
until they reach the leaves, where the elaboration is
performed. The process is carried on under the law,
that if two liquids of different density be separated by
a thin or porous diaphragm, they will permeate the
diaphragm, and change places with greater or less
rapidity according to circumstances, the liquid of *less*
density penetrating the diaphragm more rapidly than
the other. The cells of plants, as we have said, contain
dense liquid matter. The moisture present in the soil,
and in contact with the tender root-hairs (which are
made up of cells, you will remember), being of less density than the contents of the cells, flows into them, and
is then passed on from cell to cell on the same princi-
ple. The supply of assimilated matter is thus renewed
as fast as it is appropriated by the newly divided and
growing cells.

166. If a plant, during its existence, simply multiplies
its cells in this way, it can of course only be a mass of
cellular tissue as long as it lives. But we see everywhere about us plants, such as trees and shrubs, whose
stems are extremely firm and enduring. How do these
stems differ from those of tender herbs? How do they
differ from the soft parts of the plants to which they
themselves belong? A moment's consideration will
make it evident that, as every plant begins with a single
cell, and increases by successive multiplications of it,
every part of the plant must at some time have been
composed of cellular tissue, just as the newer portions
are at present. The cells of those parts which are no

longer soft must, then, have undergone a change of
some kind. Let us try to understand the nature of
this change. It has been stated that the walls of new
cells are extremely thin ; as they become older, however,
they, as a rule, increase in thickness, owing to deposits
of cellulose upon their inner surface. It sometimes
happens, indeed, that the deposits are so copious as to
almost completely fill up the cavity of the cell. The
idea will naturally suggest itself, that this thickening of
the walls must impede the passage of the sap, but it is
found that the thickening is not uniform, that there
are, in fact, regular intervals which remain thin, and
that the thin spot in one cell is directly opposite a corresponding thin spot in the wall of its neighbour.
Eventually, however, these altered cells cease to convey
sap.

167. The hard parts of plants, then, differ from the
soft parts in the different *consistency of their cell-walls.*
But they differ also in the *form of the cells
themselves.* In those parts where toughness
and strength will be required, as, for example, in the inner bark, in the stem, and
in the frame-work of the leaves, the cells
become elongated and their extremities assume a tapering form, so that they overlap
each other, instead of standing end to end as
in ordinary cellular tissue (Fig. 165). To
this drawing-out process, combined with the
hardening of the walls, is due the firmness

Fig. 165.

of wood generally, and the tissue formed by these modi-
fied cells is known as *woody tissue.* On account of
the great relative length of the cells found in the inner

bark, and the consequent toughness conferred upon that part, the tissue formed by them is specially distinguished as *bast* tissue. Associated with the wood-cells are commonly found others, differing from them chiefly in being larger in diameter, and formed out of rows of short cells, standing end to end, by the disappearance of the partitions which separated them. These enlarged cells, produced in this way, are called *vessels* or *ducts*, and a combination of them is known as *vascular tissue*. Ducts invariably show markings of some sort on their walls. The one figured in the margin (Fig. 166) is a *dotted* duct, the dots being spaces which have not been thickened by deposits of cellulose. Other ducts are *spirally* marked on their inner surface, but in this case the markings are themselves the thickened part of the cell-wall. It is convenient to speak of the mixture of woody and vascular tissue as the *fibro-vascular system*.

Fig. 166.

The name *parenchyma* is commonly applied to ordinary cellular tissue, whilst tissue formed of long cells is called *prosenchyma*.

It will be understood, then, that all cells of every description, found entering into the composition of a plant, are only modifications of one original form, the particular form ultimately assumed by the cells depending mainly on the functions to be discharged by that portion of the plant in which the cells occur.

EXOGENOUS AND ENDOGENOUS STEMS.

168. It has already been hinted that the two great classes of plants, Dicotyledons and Monocotyledons, differ in the mode of growth of their stems. We shall

now explain somewhat more in detail the nature of this difference. Bearing in mind the fact stated in the preceding part of the chapter, that old and new parts differ mainly in the shape of their component cells and the texture of the cell-walls, it will be found that the distinction between Exogenous and Endogenous growth depends mostly upon the relative situation of the new cells and the old ones—of the *parenchyma* and the *prosenchyma*.

169. Let us begin with the stem of a Dicotyledon. Fig. 167 shows a section of a young shoot. The whole of the white part is cellular tissue, the central portion being the *pith*. The dark wedge-shaped portions are fibro-vascular bundles, consisting mainly of woody tissue, a few *vessels*, easily recognised by their larger openings, being interspersed.

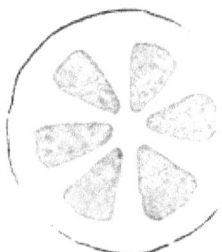

Fig. 167.

As the shoot becomes older, these bundles enlarge, and others are formed between them, so that the radiating channels of cellular tissue which separate them are in the end reduced to much smaller compass than in the earlier stages of growth (Fig. 168). The narrow channels are the *medullary rays*. The cells of which they are composed are flattened by compression. Eventually, a *ring* of wood is

Fig. 168.

formed, the medullary rays intersecting it in fine lines, as the sawed end of almost any log will show. Outside the zone of wood is the *bark*, which at first consists altogether of cellular tissue. As the season advances,

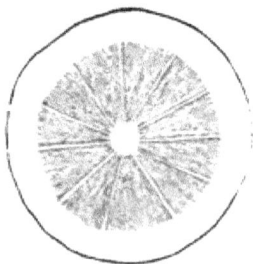

however, long *bast* cells are formed in the *inner* part, next the wood, which part is thereafter specially designated the *liber*. The outer ring of all, enclosing the whole stem, is the *epidermis* or *skin*.

170. It is now to be observed that, year after year, the rings of wood are increased in thickness *by the multiplication of their outer cells.* There is, consequently, *always* a layer of soft cells between the wood and the bark. This is known as the *cambium layer*, and it is here that the whole growth of an exogenous stem takes place. The soft cells on its inner side are gradually transformed into woody tissue and vessels, whilst those on its outer side become the bast cells of the liber, and others form the extension of the medullary rays.

Bear in mind, then, that the exogenous stem is characterized (1) by the formation of its wood in rings, (2) by the presence of the continuous cambium-layer, and (3) by the presence of a true bark.

171. Let us now consider the structure of an endogenous stem. Fig. 169 represents a section of one. Here, again, the white portion is cellular tissue, whilst the dark parts are the fibro-vascular bundles. This stem is at once distinguished from the other by the isolation of these bundles. They never coalesce to form a ring. That portion of each bundle, which is nearest the centre of the stem, corresponds to the wood of the exogen, whilst the outer portion of each consists of cells which resemble the exogenous bast-cells, but *there*

Fig. 169.

is no cambium-layer, and consequently no arrangement for the indefinite continuance of the growth of the bundles. Once formed, therefore, they remain unchanged, and the growth of the stem consists in the production of new ones. These (which originate at the bases of new leaves) being introduced amongst the older ones, act as wedges, and swell the stem as a whole.

THE FOOD OF PLANTS.

172. A word or two is necessary on this subject in addition to what has already been said. The nature of a plant's food may be determined by making a chemical analysis of the plant's substance. As already stated, the chemical elements found in plants are chiefly four, carbon, oxygen hydrogen and nitrogen, the latter element occurring in the protoplasm of active cells. What, then, are the sources from which the plant obtains these materials of its growth ? In the atmosphere there is always present a gas known as carbon dioxide, or carbonic acid. This gas, which is a compound of carbon and oxygen, is produced largely in the lungs of animals, and by them exhaled. It is readily soluble in water, so that rain-drops in their passage through the air dissolve it and carry it with them into the soil. Again, wherever animal or vegetable matter is decaying there is produced a gas called ammonia, a compound of nitrogen and hydrogen, and, like carbonic acid, readily soluble, so that this also is present in rain-water. And when it is considered that a very large proportion of the air consists of free nitrogen soluble to some extent in water and that the elements of water itself are oxygen and hydrogen it will be evident that the moisture in the earth contains a supply of every one of the element

chiefly required by the plant. Now it is a matter of common experience that, with rare exceptions, a plant will wither and die unless supplied with adequate moisture. We therefore come to the conclusion, that at any rate the greater part of the nourishment of plants is imbibed in liquid form through the roots. The law of endosmose, in accordance with which this imbibing goes on, has already been explained. The *sap*, as it is called, ascends through the newer tissues, and is attracted to the leaves by the constant evaporation going on there, and the consequent thickening of the contents of the cells in those organs.

173. And this leads to the question—How does the water-vapour make its escape from the leaves? The microscope solves this difficulty for us. A leaf almost always presents one surface towards the sky and the other towards the ground. It is protected on both sides by an epidermis or skin, consisting of very closely packed cells. The side exposed to the sun is almost unbroken, but the lower side is seen, under the microscope, to be perforated by innumerable little openings, which lead into the body of the leaf. These openings, to which the name *stomata*, or *stomates* (Fig. 170) has been given, have the power of expanding when moistened by damp air, and contracting when dry. By this wonderful contrivance, the rate of evaporation is regulated,

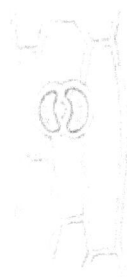

Fig. 170.

and a proper balance maintained between the supply at the root and the loss from the leaves. The stomates, it may be noticed, serve also as means whereby carbonic acid may be directly absorbed from the air. In those plants whose leaves float on water the stomates are found on the upper surface, and in vertical leaves they occur pretty equally on both surfaces. Immersed leaves are without stomates.

174. The *crude sap*, then, which ascends into the leaves is concentrated by the evaporation of its superfluous water. When so concentrated, the action of sunlight, in connection with the green colouring matter existing in the cells of the leaves, and known as chlorophyll, decomposes the carbonic acid, contained in the sap, into its carbon and oxygen. The latter gas issues from the leaves into the air, whilst the carbon is retained and combined with the remaining elements to form *elaborated sap*, out of which the substance of new cells is constructed.

175. It thus appears that the chemical action which goes on in the leaves of plants is precisely the reverse of what takes place in the lungs of animals. The latter *inhale* oxygen, combine it with the carbon of the blood, and *exhale* the resulting carbonic acid. The former take in carbonic acid, decompose it in the leaves and other green parts, and *exhale* the oxygen. Plants may therefore be regarded as purifiers of the air.

176. It remains to be added, that besides the four substances, carbon, oxygen, hydrogen, and nitrogen, which are called the *organic elements*, many others are found in the fabric of plants. When a piece of wood is burnt away, the organic elements disappear, but a quantity of *ash* remains behind. This contains the various mineral substances which the water absorbed by the plant has previously dissolved out of the soil, but which it is not necessary to our present purpose to enumerate here.

CHAPTER XVI.

CLASSIFICATION OF PLANTS ACCORDING TO THE NATURAL SYSTEM.

177. Hitherto, our examination of plants has been confined to a few selected specimens, and we have examined these chiefly in order to become acquainted with some variations in the details of growth, as exemplified by them. Thus, we have found plants which agree in exhibiting two cotyledons in the embryo, and others, again, which are monocotyledonous. Some members of the former group were found to exhibit two sets of floral envelopes, other only one, and others, again, were entirely without these organs. And so on through the various details. We now set out with the vegetable world before us—a world populated by forms almost infinite in number and variety. If, therefore, our study of these forms is to be carried on to advantage, we shall have to resolve upon some definite plan or system upon which to proceed; otherwise we shall merely dissipate our energies, and our results will be without meaning. Just as, in our study of language, we find it convenient to classify words into what we call parts of speech, and to divide and subdivide these again, in order to draw finer distinctions, so, in our study of plants, it will be necessary to arrange them

first of all into comprehensive groups, on the ground of some characteristic possessed by every member of each group. Just as, in Latin, every noun whose genitive case is found to end in æ is classed with nouns of the first declension, so in Botany, every plant presenting certain peculiarities will be placed in a group along with all the other plants presenting the same peculiarities.

17C. Some hints have already been given you as to the kind of resemblances upon which classification is based. For instance, an immense number of plants are found to produce seeds with a dicotyledonous embryo, while an immense number of others have monocotyledonous embryos. This distinction, therefore, is so pronounced, that it forms the basis of a division into two very large groups. Again, a very large number of dicotyledonous plants have their corollas in separate petals ; many others have them united, whilst others again have no petals at all. Here, then, is an opportunity to subdivide the Dicotyledons into polypetalous, gamopetalous, and apetalous groups. And so we go on, always on the plan that the more widely spread a peculiarity is found to be, the more comprehensive must be the group based on that peculiarity ; and so it happens, that the smallest groups of all come to depend upon distinctions which are, in many cases, by no means evident, and upon which botanists often find themselves unable to agree.

179. As our divisions and subdivisions will necessarily be somewhat numerous, we shall have to devise a special name for each kind of group, in order to avoid confusion of ideas. We shall, then, to begin with, draw a broad line of distinction between those plants

which *produce flowers* of some kind, and those which *do not*, and to each of these great groups we shall give the name **Series**. We thus have the **Flowering**, or, to use the Greek term, Phanerogamous, **Series**, and the **Flowerless**, or Cryptogamous, Series; or we may speak of them briefly as **Phanerogams** and **Crypto- gams**. Then, leaving the Cryptogams aside for the moment, we may break up the Phanerogams into two great **Classes**, **Exogens** (or D:cotyledons) and **Endogens** (or **Monocotyledons**), for reasons already explained. By far the greater number of Exogens produce seeds which are enclosed in a pericarp of some kind; but there is a remarkable group of plants (repre- sented in Canada only by the Pines and their imme- diate relatives) which dispense with the pericarp alto- gether, and whose seeds are consequently naked. So that we can make two Sub-Classes of the Exogens, on the basis of this difference, and these we shall call the **Angiospermous Sub-Class**, and the **Gymnosper- mous** (naked-seeded) **Sub-Class**. The first of these may be grouped in three **Divisions**, the *Polypetalous*, *Gamopetalous*, and *Apetalous*, and the **Endogens** also in three, the *Spadiceous*, the *Petaloideous*, and the *Glumaceous*, types of which we have already examined in the Marsh Calla (spadiceous), Trillium (petaloideous), and Timothy (glumaceous), and the distinctions between which are sufficiently obvious.

The Cryptogams are divided into three great **Classes**, viz.: Acrogens, embracing **Ferns**, **Horse- tails** and **Club-mosses**; Anophytes, embracing **Mosses** and Liverworts; and Thallophytes, embracing Lichens, Seaweeds, and Mushrooms.

So far, then, our classification is as follows:

VEGETABLE KINGDOM.	Series I. Phanerogams	Class I.—Exogens ...	Sub-class 1—Angiosperms
			Polypetalous Division.
			Gamopetalous
			Apetalous
			Sub-class 2-Gymnosperms
		Class II.—Endogens	Spadiceous Division.
			Petaloideous Division.
			Glumaceous Division.
	Series II. Cryptogams.	Class III.- Acrogens.	
		Class IV.—Anophytes.	
		Class V.—Thallophytes.	

Each of the *Divisions* is sub-divided into a number of Families or Orders ; each Order into a number of Genera ; and each Genus into Species. A species is the sum of all the individual plants whose resemblances in all essential respects are so great as to warrant the belief that they have sprung from one common stock. De Candolle has this statement : " We unite under the designation of a *species* all those individuals that mutually bear to each other so close a resemblance as to allow of our supposing that they may have proceeded originally from a single being or a single pair." We may also speak of each one of these individual plants as a species. For example, you may say, after finishing the first lesson of this book, that you have examined *a species of Buttercup.* Mere differences-of colour or size are not sufficient to constitute different species. The Balsams of our gardens, for instance, are of various colours, and the plants vary greatly in size, yet they all belong to one species. These minor differences, which are mainly the result of care and cultivation, give rise to *varieties.* These are of great interest to the horticulturist, but the study of *species* is the great end and aim of the botanist.

180. Those Species which are considered to resemble each other most nearly are grouped into Genera, and the Genera, in like manner, into Orders; but these particular groupings are more or less artificial, and are subject to continual alteration in consequence of our imperfect knowledge. As, year by year, new facts are brought to light, modifications in arrangement take place. In the Classification which constitutes the Second Part of this work, the Divisions spoken of above are placed in the order named. In the Polypetalous Division, those Orders are put first which embrace plants with *hypogynous stamens* and *apocarpous pistils*, the parts of the flowers being consequently separate; then those with similarly inserted stamens, but *syncarpous pistils;* then those with *perigynous* stamens; and, generally, we proceed from plants whose flowers have all their parts separate to those exhibiting more or less *cohesion* and *adhesion*, and finally to those having one or more parts of the flower wanting.

181. In looking up the name of a plant, it will be your object to determine the *Genus* to which it belongs, and also the *Species*. The name of every plant consists of two parts: its Genus first, and then its Species. The name of the Genus is a Latin noun, and that of the Species a Latin adjective agreeing with the noun. The Buttercup, for example, which we examined at the outset, belongs to the Genus *Ranunculus*. In this Genus are included many Species. The particular one examined by us is known as *acris ;* so that the full name

of the plant is *Ranunculus acris.* In like manner, the name of the plant popularly called Marsh-Marigold is *Caltha palustris.*

182. The *Key* which is prefixed to the Classification will enable you to determine without much difficulty the *Order* to which a plant belongs, but nothing more. Having satisfied yourselves as to the Order, you must turn to the page on which that Order is described, and, by carefully comparing the descriptions there given with the characters exhibited by your plant, decide upon its Genus, and, in the same manner, upon its Species.

THE HERBARIUM.

Those who are anxious to make the most of their botanical studies will find it of great advantage to gather and preserve specimens for reference. A few hints, therefore, on this subject will not be out of place. It will, of course, be an object to collectors to have their specimens exhibit as many of their natural characters as possible, so that, although dried and pressed, there will be no difficulty in recognizing them ; and to this end neatness and care are the first requisites.

Specimens should be collected when the plants are in flower, and, if possible, on a dry day, as the flowers are then in better condition than if wet. If the plant is small, the whole of it, root and all, should be taken up ; if too large to be treated in this way, a flower and one or two of the leaves (radical as well as cauline, if these be different,) may be gathered.

As many of your specimens will be collected at a distance from home, a close tin box, which may be slung over the shoulder by a strap, should be provided, in which the plants may be kept fresh, particularly if a few drops of water be sprinkled upon them. Perhaps a better way, however, is to carry a portfolio of convenient size—say 15 inches by 10 inches—made of two pieces of stout pasteboard or thin deal, and having a couple of straps with buckles for fastening it together. Between the covers should be placed sheets of blotting paper, or coarse wrapping paper, as many as will allow the specimens to be separated by at least five or six sheets. The advantage of the portfolio is, that the

plants may be placed between the sheets of blotting
paper, and subjected to pressure by means of the straps,
as soon as they are gathered. If carried in a box, they
should be transferred to paper as soon as possible. The
specimens should be spread out with great care, and the
crumpling and doubling of leaves guarded against. The
only way to prevent moulding is to place plenty of
paper between the plants, and *change the paper frequently,*
the frequency depending on the amount of moisture
contained in the specimens. From ten days to a fort-
night will be found sufficient for the thorough drying
of almost any plant you are likely to meet with. Having made a pile of specimens with paper between them,
as directed, they should be placed on a table or floor,
covered by a flat board, and subjected to pressure by
placing weights on the top ; twenty bricks or so will
answer very well.

When the specimens are thoroughly dry, the next
thing is to mount them, and for this purpose you will
require sheets of strong white paper ; a good quality of
unruled foolscap, or cheap drawing paper, will be suitable. The most convenient way of attaching the specimen to the paper is to take a sheet of the same size as
your paper lay the specimen carefully in the centre,
wrong side up, and gum it thoroughly with a very soft
brush. Then take the paper to which the plant is to
be attached, and lay it carefully on the specimen. You
can then lift paper and specimen together, and, by
pressing lightly with a soft cloth, ensure complete adhesion. To render plants with stout stems additionally
secure, make a slit with a penknife through the paper
immediately underneath the stem ; then pass a narrow
band of paper round the stem, and thrust both ends of

the band through the slit. The ends may then be gummed to the back of the sheet.

The specimen having been duly mounted, its botanical name should be written neatly in the lower right-hand corner, together with the date of its collection, and the locality where found. Of course only one Species should be mounted on each sheet; and when a sufficient number have been prepared, the Species of the same Genus should be placed in a sheet of larger and coarser paper than that on which the specimens are mounted, and the name of the Genus should be written outside on the lower corner. Then the Genera of the same Order should be collected in the same manner, and the name of the Order written outside as before. The Orders may then be arranged in accordance with the classification you may be using, and carefully laid away in a dry place. If a cabinet, with shelves or drawers, can be specially devoted to storing the plants, so much the better.

INDEX AND GLOSSARY

The references are to the Sections, unless Figures are specified.

Apocarpous: applied to pistils when the carpels are free from each ether.

Appendage: anything attached or added.

Appressed: in contact, but not united.

Aquatic: growing in the water, whether completely, or only partially, immersed.

Arborescent: resembling a tree.

Aril, 157.

Arrow-shaped, Fig. 120.

Ascending: rising upward in a slanting direction; applied chiefly to weak stems.

Ascending axis: the stem of a plant.

Ascidium: a pitcher-shaped leaf, Fig. 134.

Ashes of plants, 176.

Assimilation, 165.

Auriculate: same as *auricled*, having rounded lobes at the base; applied mostly to leaves.

Awl-shaped, Fig. 112.

Awn: a bristle, such as is found on the glumes of many Grasses, Barley for example.

Axil, 3.

Axile: relating to the axis.

Axillary: proceeding from an axil.

Axillary buds, 88.

Axillary flowers, 120.

Axis: the stem and root.

baccate: like a berry.

Bark, 169.

Bast, 167.

Bearded: furnished with hairs, like the petals of some Violets, &c.

Bell-shaped, 130.

Berry, 149.

Biennial: a plant which grows from seed in one season, but produces its seed and dies in the following season.

Bifoliolate: having two leaflets.

Bilabiate: two-lipped, Fig. 142.

Bipinnate: twice pinnate, Fig. 132.

Bipinnatifid: twice pinnatifid, Fig. 123.

Blade: the broad part of a leaf or petal.

Bracts, 19, 125.

Bracteate: subtended by a bract.

Bractlets: secondary bracts growing on pedicels, 125.

Branches: growths from the sides of a stem, originating in axillary buds, 3.

Breathing-pores (stomates), 173.

Bud: an undeveloped stem or branch.

Bulb, 92.

Bulbiferous : producing bulbs.
Bulblets, 95.
Bulbous : like a bulb in shape.

Caducous, **128.**
Calyx, 5.
Cambium layer, 170.
Campanulate, 130.
Capillary : fine and hair-like.
Capitulum : same as *head*, 122.
Capsule, 155.
Carina, or keel: the two coherent petals in the front of a flower
    of the Pea kind, Fig. 36.
Caryopsis, 156.
Carpel, 7.
Carpellary : relating to a carpel, *e.g.*, a *carpellary leaf*, &c.
Cartilaginous : tough.
Catkin, Figs. 63, 64.
Caulescent : with an evident stem.
Caulicle : another name for the radicle.
Cauline : relating to the stem, *e.g.*, *cauline leaves*, &c., **4.**
Cell : the hollow in the anther, which contains the pollen.   See
    also 161.
Cell-multiplication, 164.
Cellular tissue, 162.
Cellulose, 163.
Centrifugal inflorescence, **121.**
Centripetal inflorescence, **120.**
Chalaza : the **part** of an **ovule** where the coats **are united to**
    the nucleus.
Chlorophyll, 163, 174.
Ciliate, 116.
Circinate : curled up like the young frond of a Fern.
Circulation in cells, 163.
Circumcissile: opening like a pyxis, Fig. 161.
Classification, 177.
Claw (of a petal), 40, 129.
Climbing stems, 90.
Club-shaped : with the lower part more slender than the upper,
    as the style of Dog's-tooth Violet, Fig. 73.
Cohesent : a term applied to the union of like parts, 26.
Cohesion, 26.
Collerm, or neck : the junction of the stem and root.
Collactive fruits, 153.
Column, 72.
Coma : a tuft of hairs, such as that on the seed of Dandelion,
    Fig. 56.
Complete, 8.
Compound, or Composite, flowers, **49.**

Compound leaf, 100.
Compound spike, corymb, &c., 122.
Cone, 142.
Coniferous : bearing cones.
Connate : grown together.
Connate-perfoliate, Fig. 130.
Connective, 132.
Convolute : rolled inwards from one edge.
Cordate, 108.
Corm, 66.
Corolla, 5.
Corymb, Fig. 135.
Corymbose : like a corymb.
Cotyledons, 58.
Creeping, 90.
Crenate, Fig. 128.
Cruciform : cross-shaped, as the flowers of Shepherd's Purse, &c.
Crude sap, 174.
Cryptogams, 179.
Culm, 90.
Cuneate : wedge-shaped.
Cuspidate, Fig. 126.
Cyclosis, 163.
Cyme, 124.
Cymose : like a cyme.

Decandrous : with ten separate stamens.
Deciduous 5.
Decompoun : applied to leaves whose blades are divided and
    subdivided.
Decumbent : applied to stems which lie on the ground but turn
    upward at the extremity.
Decurrent, Fig. 131.
Decussate : applied to the arrangement of leaves, when successive
    pairs of opposite leaves are at right angles, as in the plants
    of the Mint family.
Definite inflorescence, 121.
Deflexed : bent down.
Dehiscent, 147.
Dehiscence of anthers, Figs. 147, 148, 149.
Deliquescent : applied to stems which dissolve into branches.
Deltoid, 146.
Dentate, 112.
Depauperate : unnaturally small.
Depressed : flattened down.
Descending axis : the root, 83.
Determinate inflorescence, 121.
Diadelphous : applied to stamens, 66.
Diandrous : with two separate stamens.

Dichlamydeous: having both sets of floral envelopes.
Dicotyledonous, 58.
Dicotyledons, 59.
Didynamous (stamens). 50
Digitate, 101.
Diœcious, 56.
Disk: in flowers of the Composite Family, the centre of the head as distinguished from the border ; a fleshy enlargement of the receptacle of a flower.
Dissected: finely cut.
Dissepiment, 139.
Distinct: not coherent, (see Coherent).
Divergent : separating from one another.
Dodecandrous: with twelve distinct stamens.
Dorsal suture, 138.
Dotted ducts, Fig. 166.
Double flowers : abnormal flowers in which stamens and carpels have been transformed into petals.
Downy : covered with soft hairs.
Drupe, 147.
Drupelet, a little drupe.
Ducts, 167.

Earthy constituents of plants, 176.
Elaborated sap, 174.
Elementary constituents of plants, 176.
Elementary structure, 160.
Elliptical : same as oval, 105.
Emarginate, 111.
Embryo, 12.
Embryo.sac, 16.
Emersed : raised above the surface of water.
Endocarp : " When the walls of a pericarp form two or more layers f dissimilar texture, the outer layer is called the Epicarp, the middle one Mesocarp, and the innermost Endocarp."— Gray.
Endogen, 81.
Endogenous growth, 171.
Endosmose, 172, 165.
Enneandrous : with nine distinct stamens.
Entire, 112.
Ephemera : last one day only.
Epicalyx, 33.
Epicarp : see Endocarp.
Epidermis, 160.
Epigynous : inserted on the ovary, 46.
Epipetalous : inserted on the corolla, 47.
Epiphytes, 87.
Equitant (leaves), 98.

Genera: plural of genus.
Genus, 179.
Germ : same as embryo.
Germination, 150.
Gibbous : swollen on one side.
Glabrous, 116.
Gladiate : sword-shaped.
Glands: applied generally to cells or hairs on the surfaces of
    plants, in which resinous or oily matters are secreted ; but
    the term is also used to describe any projection, the use of
    which is not clear.
Glandular : bearing glands.
Glaucous, 116.
Globose: like a globe or sphere.
Glumaceous : bearing or resembling glumes.
Glumes, 75.
Gourd, 149.
Grain, 156.
Granules : particles.
Gymnospermous, 143.
Gymnosperms, 179.
Gynœcium, 137.
Gynandrous, 135.

Habitat : a term applied to the region most favourable to the
    growth of a plant : the place where it grows naturally.
Hairs, 116.
Hairy, 4.
Halberd-shaped, Fig. 119.
Hastate. Fig. 119.
Head, 122.
Heart-shaped, 108.
Heptandrous: with seven distinct stamens.
Herb, 89.
Herbaceous, 89.
Herbarium : a botanist's collection of dried plants.
Hexandrous: with six distinct stamens.
Hilum, 157.
Hirsute : rough with hairs.
Hispid : covered with stiff hairs.
Hoary : densely covered with fine grayish hairs
Hortus siccus : same as herbarium.
Hybrids: plants resulting from the crossing o. nearly related
    species.
Hypogynous, 135.

Imbricate : overlapping like shingles on a roof.
Immersed : wholly under water.
Imperfect, 53.

Included, 136.
Incomplete, 19.
Incurved (petals) Fig. 50.
Indefinite, 26, 134.
Indefinite inflorescence, 120.
Indehiscent, 147.
Indeterminate inflorescence, 120.
Indigenous : naturally growing in a country.
Inferior : underneath ; farthest from the axis ; the ovary is
    inferior when the calyx adheres to it throughout ; the calyx
    is inferior when free from the ovary.
Inflorescence, 119.
Innate, 132.
Inserted : attached to.
Insertion : the point, or manner, of attachment.
Internodes, 4.
Interruptedly pinnate, Fig. 133.
Introrse, 132.
Involucel, 125.
Involucre, 125.
Involute : rolled inwards from both edges.
Irregular, 35.
Isomerous : having the parts equal in number.

Joints : a name sometimes given to the nodes of a stem.

Keel, see Carina.
Kernel, 16.
Key-fruit, 156.
Kidney-shaped, Fig. 121.

Labellum (or lip), 71.
Labiate, 50.
Lanceolate, Fig. 113.
Leaf, 97.
Leaf-arrangement, 99.
Leaf-green, see Chlorophyll.
Leaflet, 100.
Leafstalk, 4.
Legume, 154.
Leguminous : producing or relating to legumes.
Liber, 169.
Ligneous : woody.
Ligulate, 131.
Ligule : a strap-shaped corolla   n Grasses, a scale-like projec-
    tion between the blade of a leaf and the sheath.
Limb, 129, 130.
Lip, see Labellum.
Linear, Fig. 111.

Racemose: like a raceme.
Rachis: an axis.
Radiate, 101.
Radical: pertaining to the root.
Radical leaves, 4.
Radicle, 58.
Raphides, 163.
Ray : the marginal florets of a composite flower, as distinguished from the disk.
Receptacle, 8.
Recurved : curved backwards.
Reflexed : bent backwards.
Regular : with parts of the same size and shape.
Reniform, Fig. 121.
Reticulated : netted.
Retuse : slightly notched at the apex.
Revolute : rolled back.
Rhizome, 91.
Ribs, 101.
Ringent, 131.
Root, 2, 83.
Root-hairs, 165.
Rootlet, 2.
Rootstock, 91.
Rotate, 130.
Rotation in cells, 163.
Rudimentary : imperfectly developed.
Rugose : wrinkled.
Runcinate: with teeth pointing backwards, as in the leaf Dandelion.
Runner, 90.

Sagittate, Fig. 120.
Salver-shaped, Fig. 141.
Samara, Fig. 162.
Sap, 172, 174.
Sarcocarp : the flesh of a drupe.
Scabrous : rough.
Scandent : climbing.
Scape, 19.
Scar, 157.
Scion : a young shoot.
Seed, 17, 157, 158.
Seed-vessel, see Ovary.
Sepal, 5.
Septicidal (dehiscence): splitting open along th partitions.e
Septum : a partition.
Series, 179.
Serrate, 112

Vascular tissue, 167.
Veins : the finer parts of the framework of a leaf.
Venation, 101.
Ventral suture, 138.
Vernation, same as Præfoliation.
Versatile, 132.
Vertical leaves, 98.
Verticillate, 99.
Vessels, 167.
Villose, 116.

Wavy : with alternate rounded hollows and projections, 112.
Wedge-shaped : like a wedge, the broad part being the apex.
Wheel-shaped, see Rotate.
Whorl : a circle of three or more leaves at the same node.
Woody tissue, 167.

APPENDIX.

Selections from Examination Papers.

UNIVERSITY OF TORONTO.

1. Define suckers, stolons, offsets, runners, tendrils, thorns, and prickles, describing their respective origins and uses, and giving examples of plants in which they occur

2. What are the functions of leaves? Describe the different kinds of compound leaves.

3. What is meant by inflorescence? Describe the different kinds of flower-clusters, giving an example of each.

4. Mention and explain the terms applied to the various modes of insertion of stamens.

5. How are fruits classified? What are multiple or collective fruits? Give examples.

6. Relate the differences in structure between endogenous and exogenous stems. Describe their respective modes of growth.

7. What is the food of plants? how do they obtain it? and how do they make use of it?

8. Describe the component parts of a simple flower. How is reproduction effected?

9. Describe the anatomical structure of a leaf, and the formation and office of leaf-stomata.

10. Explain the consequences of flowering upon the health of a plant, and shew how these effects are remedied in different climates. What practical bearing has this upon horticulture?

11. Trace the development of a carpel from a leaf. Describe the different forms assumed by placentæ in

compound ovaries, and explain the origin of these variations.

12. Mention the principal modes in which pollen gains access to the stigma. What are hybrid plants, and how are they perpetuated?

13. Describe the anatomy of a leaf. What are stomata?

14. What is the placenta in a seed vessel? Describe the different modes of placentation. Shew how the varieties of placentation agree with the "altered-leaf theory" of the pistil.

15. Give the characters of the Compositæ. How is the order subdivided? Describe the composite flower, and mention some of the common Canadian examples of this order.

16. Give the peculiarities of Endogens in seed-leaf, leaf, and stem. Subdivide the class. Describe shortly the orders Araceæ and Gramineæ.

17. Describe the wall of a seed-vessel, and notice its varieties of form.

18. What is meant by the dehiscence of a capsule? Shew the different modes in which pods dehisce, and give examples of each.

19. Give the characters and orders of Gymnospernous Exogens.

20. Give the characters of Ranunculaceæ. Describe shortly some of the principal plants of the order.

21. Give some account of the *special forms* which the leaves of plants assume.

22. What are stipules? What their size and shape?

23. What is meant by Imperfect, Incomplete, and Unsymmetrical flowers respectively?

24. Describe Papilionaceous and Labiate corollas.

25. Write notes on Abortive Organs, Gymnospermous Pistil, and Pollen Granule.

26. Distinguish between the essential and non-essential materials found in plants, and notice the non-essential.

27. What is vegetable growth? Illustrate by a ref-

erence to the pollen granule in its fertilization of the ovary.

28. What is an axil ?   What is the pappus ?

29. What are the cotyledons ?   What is their function, and what their value in systematic Botany ?

30. Distinguish between Epiphytes and Parasites; Describe their respective modes of growth, and give examples of each.

31. What is the difference between roots and subterranean branches ?   Define rhizoma, tuber, corm, and bulb, giving examples of each. How does a potato differ botanically from a sweet-potato ?

32. Describe the calyx and corolla ; what modifications of parts take place in double flowers ?

33. What is a fruit in Botany ?   Explain the structure of an apple, grape, almond, strawberry, fig, and pineapple.

34. What organs appear in the more perfect plants ? In what two divisions are they comprised ?

35. Weak climbing stems distinguished according to the mode in which they support themselves, the direction of their growth, the nature of their clasping organs.

36. Structure and parts of a leaf : What is most important in their study ? Give the leading divisions, and mention what secondary distinctions are required in specific description ?

37. Function of the flower : its origin : its essential and accessory parts : names of the circles and their component organs : circumstances which explain the differences among flowers.

38. Parts of the fully formed ovule and distinctions founded on their relative position.

39. Sub-kingdoms and classes of the vegetable kingdom.

SECOND CLASS TEACHERS' CERTIFICATES, PROVINCE OF ONTARIO.

1. Name the parts of the pistil and stamens of a flower and give their uses

2. What are Perennial plants? Describe their mode of life.

3. "There are two great classes of stems, which differ in the way the woody part is arranged in the cellular tissue." Fully explain this.

4. Describe the functions of leaves. How are leaves classified as to their veining?

5. Name and describe the organic constituents of plants.

6. Name the organs of reproduction in plants, and describe their functions.

7. Give, and fully describe, the principal parts of the flower.

8. What are the different parts of a plant? Describe the functions of each part.

9. State all the ways by which an Exogenous stem may be distinguished from an Endogenous.

10. Describe the functions of leaves. What is the cause of their fall in autumn? Draw and describe a maple leaf.

11. Name the different parts of a flower, and describe the use of each part. Draw a diagram showing a stamen and a pistil and the parts of each.

12. What is the fruit? Why do some fruits fall from the stem more easily than others?

13. Of what does the food of plants consist? In what forms and by what organs is it taken up, and how is it asssimilated? Name the substances inhaled and those exhaled by plants, and the uses of each in the economy of nature.

14. Describe fully (1) the plant in Vegetation; (2) the plant in Reproduction.

15. Describe Fibrous roots, Fleshy roots, and different kinds of Tap-root.

16. Describe the structure and veining of leaves.

17. "The nourishment which the mother plant provides in the seed is not always stored up in the embryo." Explain and illustrate.

18. Describe the various modes in which Perennials

"provide a stock of nourishment to begin the new growth."

19. Describe fully the organs of reproduction in a plant. Describe the process of germination.

20. What are the parts of a flower? Give illustrations by diagram, with a full description.

21. Name and describe the principal sorts of flowers.

22. What elementary substances should the soil contain for the nourishment of plants?

23. How are plants nourished before and after appearing above ground?

24. Tell what you know about the various forms of the calyx and the corolla.

25. Explain the terms Cotyledon, Pinnate, Rootstock, Filament, and Radicle.

26. Explain the terms Papilionaceous, Cruciferous, Silique, and Syngenesious; and in each case name a family in the description of which the term under consideration may be properly applied.

27. Give the characters of the Rose family.

28. Describe the various modes in which biennials store up nourishment during their first season.

29. Explain the meaning of the terms Sepal, Bract, Raceme, and Stipule. Describe minutely the Stamen and the Pistil, and give the names applied to their parts.

30. Are the portions of the onion, the potato, and the turnip which are capable of preservation through the winter, equally entitled to the name of roots? Give reasons for your answer.

FIRST CLASS CERTIFICATES.

1. What are the cotyledons? Describe their functions, &c. State their value in systematic botany.

2. Describe the difference in structure and modes of growth of exogenous and endogenous stems.

3. Describe the circulation in plants. "In the act of making vegetable matter, plants purify the air for animals." Explain this fully.

4. What are Phœnogamous plants ? Define Raceme, Corymb, Head, Panicle, Ament.

5. Give the characters of *(a)* The classes Exogens and Endogens ; *(b)* The Mint and Lily families.

6. To what family do the Cedar, Clover. Mustard, and Dandelion respectively belong ?

7. Why does a botanist consider the tuber of the potato an underground stem ?

8. Give the philosophical explanation of the nature of a flower considered as to the origin and correspondences of its different parts.

9. Draw a spathulate, an obcordate, a truncate, a palmately-divided and an odd pinnate leaf.

10. Explain the constitution of a pome or apple-fruit.

11. What organs appear in the more perfect plants, and in what divisions are they comprised ?

12. Give the function of the flower, its origin, and its essential and accessory parts.

13. Describe the nature and chief varieties of roots, and distinguish between them and underground stems.

14. "As to the Apex or Point leaves are Pointed, Acute, Obtuse, Truncate, Retuse, Emarginate, Obcordate, Cuspidate, Mucronate." Sketch these different forms.

15. "There is no separate set of vessels, and no open tubes for the sap to rise through in an unbroken stream, in the way people generally suppose." Comment on this passage.

16. The great series of Flowering Plants is divided into two *classes*. Describe these classes.

17. Give the chief characteristics of the order *crucifera (*Cress Family*),* and name some common examples of this order.

18. State the difference between definite and indefinite inflorescence, and give examples of the latter.

19. Of what does the food of plants consist ? in what form is it found in the soil ? How is it introduced into the plant ? What inference may be drawn respecting the culture of the plant ?

20. Distinguish weak climbing stems according to the mode in which they support themselves, the direction of their growth, and the nature of their clasping organs.

21. Name the three classes of Flowerless Plants, and give an example of each.

22. Explain the terms Spore, Capsule, Bract, Stipule, Albumen, and Epiphyte.

23. What are tendrils, and of what organs are they supposed to be modifications?

24. Give the characters of the Cress Family, and name as many plants belonging to it as you can.

25. Tell what you know about the minute structure and the chemical composition of vegetable tissue.

26. Describe the origin of the different kinds of placentas; and of the different parts of the fruit of the plum, the oak, and the maple.

27. Describe fully the process by which it is supposed that water is carried up from the roots of plants.

28. Give the meaning of the terms *stomate, indehiscent, thyrse, glume, pyxis.* Distinguish *epiphytes* from *parasites.*

29. Describe any plant you have examined; if you can, tabulate your description.

30. Name all the families of monopetalous dicotyledons which you remember, and give the characters of any one of them.

McGILL UNIVERSITY.

1. Describe the germination of a plant.

2. Explain the differences in the structure of the embryo.

3. Explain the functions of the Root.

4. Describe the structures in a leaf, and explain their action on the air.

5. Mention the several parts of the stamen and the pistil, and explain their uses.

6. Describe an Achene, a Samara, a Drupe and a Silique.

7. Describe the differences in the stems of Exogens and Endogens, and the relations of these to the other parts of the plant and to classification.

8. Explain the terms Genera, Species, Order.

9. What is an excurrent stem, an axillary bud, bud scales ?

10. Explain the terms *primoraial utricle, parenchyma, protoplasm,* as used in Botany.

11. What are the functions of the nucleus in a living cell?

12. Explain the movements of the sap in plants.

13. Describe the appearance under the microscope of *raphides, spiral vessels,* and *disc-bearing wood-cells.*

14. Describe the structure of the bark of an Exogen.

15. Describe freely the anatomy of a leaf.

16. Describe shortly the parts and structures denoted by the following terms, *spine, aerial root, phyllodium, cambium, stipule, rhizoma.*

17. Give examples of *phænogams, cryptogams, exogens* and *endogens,* properly arranged.

18. Describe the principal forms of indeterminate inflorescence.

19. In what natural families do we find *siliques, didynamous stamens, labiate corollas,* or *pappus-bearing achenes.* Describe these structures.

20. State the characters of any Canadian Exogenous Order, with examples.

21. Describe the cell-walls in a living parenchymatous cell.

22. Describe the fibro-vascular tissues in an Exogenous stem.

23. Describe the appearance of stomata and glandular heirs under the microscope.

24. Define *prosenchyma, corm, cyclosis, thallus.*

25. Explain the sources of the Carbon and Nitrogen of the plant, and the mode of their assimilation.

26. Describe the pericarp, stating its normal structure, and naming some of its modifications.

27. Explain the natural system in Botany, and state the gradation of groups from the species upward, with examples.

ONTARIO COLLEGE OF PHARMACY.

1. What do plants feed upon ?

2. What do you understand by the terms Acaulescent, Apetalous, Suffrutescent, Culm ?

3. Name some of the different forms of Primary, Secondary, and Aerial Roots, giving examples.

4. Explain the following terms descriptive of forms of leaves, giving sketch :—Ovate, Peltate, Crenate, Serrate, Cleft, Entire, Cuspidate, Perfoliate.

5. Explain difference between Determinate and Indeterminate inflorescence, giving three examples of each.

6. What organs are deficient in a sterile and a fertile flower ?

7. Give the parts of a perfect flower, with their relative position.

8. Give the difference between simple and compound Pistil, with example of each.

9. Name the principal sorts of buds, and explain how the position of these affects the arrangement of branches.

10. Give description of multiple and primary roots, with two examples of same ; also explain the difference between these and secondary roots.

11. Name the principal kinds of subterranean stems and branches, and explain how you would distinguish between these and roots.

12. In the classification of plants explain difference between classes and orders : genus and species.

13. Name three principal kinds of simple fruit.

14. When roots stop growing does the absorption of moisture increase or decrease ?   Give reason for it.

15. Upon what do plants live ?   Indicate how you would prove your answer correct.

16. In what part of the plant, and when, is the work of assimilation carried on ?   .

17. Name three principal kinds of *determinate*, and some of *indeterminate* inflorescence ; name the essential organs of a flower.

18. In what respects do plants differ from inorganic matter? And from animals?

19. Describe a Rhizome, Tuber, Bulb; and say if they belong to the root or stem; which are Rheum, Jalapa, Sweet Potato, Onion?

20. Define the difference between natural and special forms of leaves; between simple and compound leaves. Give example of each. Sketch a connate-perfoliate leaf.

21. Mention the parts of an embryo. Of a leaf. Of a pistil. Of a stamen. Of a seed.

22. What is meant by an albuminous seed? By diœcious flowers? By a compound ovary?

23. What is the difference between determinate and indeterminate inflorescence? How do they influence growth of the stem? Give three principal kinds of each.

24. Name the parts of a flower. What office is performed by the ovule? Name two kinds.

25. Name the parts of a vegetable cell. What are spiral ducts?

26. In what parts of the plant is the work of absorption carried on? In what part the work of assimilation? How do plants purify the air for animals?

27. Explain the natural system of classification in Botany? Name and characterize the classes of plants.

28. Explain the structure and functions of the Leaf, Bud, Root.

29. Give some of the terms used in describing the shape of a simple leaf as concerns (a) its general contour, (b) its base, (c) its margin, (d) its apex.

30. Name the organs in a perfect flower; describe fully the structure of the anther and pollen. What is coalescence and adnation of the parts of a flower?

31. Explain the terms Raceme, Pappus, Coma, Rhizome, Pentastichous.

32. State the distinction between Exogens and Endogens.

33. What are cellular structures as distinguished from vascular? What is chlorophyll?

34. Mention the organs of fructification, and explain the process of fertilization in a flowering plant.

35. Explain the structure of a seed, and describe in a few words the process of germination.

36. Define what is meant by the following terms:— Morphology, Polycotyledonous, Epiphyte, Peduncle, Stipules.

37. Describe briefly the root stem, leaf, and flower of the common dandelion, giving the functions or office of each.

38. Name some of the most common forms of leaves, giving a few rough outlines.

39. Of what part of the flower does the fruit nominally consist ? What additional parts are in some instances present ?

40. Define the terms Drupe, Pollen, Gynandrous, Pome, Adnate.

41. Explain the process of fertilization in flowering plants, and mention the different ways in which it is brought about.

49. Enumerate the different varieties of tissue recognized by botanists, and give their situation in an Endogenous stem.

THE END.

THE

COMMONLY OCCURRING

WILD PLANTS OF CANADA,

AND MORE ESPECIALLY OF

THE PROVINCE OF ONTARIO.

A FLORA FOR THE USE OF BEGINNERS.

BY

H. B. SPOTTON, M.A., F.L.S.,

PRIN. OF BARRIE COLL. INST.,

Author of "The Elements of Structural Botany."

SECOND EDITION

W. J. GAGE & COMPANY.

CONTENTS.

PREFACE.

A few words will not be out of place by way of preface to the List of Common Canadian Plants contained in the following pages. It will be observed that the List is confined to wild plants, the exclusion of cultivated Species having been determined on, partly because of the difficulty of knowing where to stop, when an enumeration of them has once been entered upon, and partly because it was thought that, on the whole, more important results would be attained by directing attention exclusively to the denizens of our own woods and fields. What is really desired is, to create among our young people an interest in the Botany of Canada, and it seems not unreasonable to hope that this end may be better attained by placing within their reach some such handy volume as the present, dealing only with such plants as grow spontaneously within our limits.

The great majority of the plants described have been personally examined, and their occurrence verified, by the writer, his observations having been directed to what may fairly be regarded as representative districts of the older Provinces, but special acknowledgments are also due to Prof. Macoun, of the Geological Survey, for the free use of his valuable notes, and other friendly assistance.

Whilst diligence has been exercised that no commonly occurring plant should be omitted, yet it can hardly be that such omissions do not occur, and the writer will be most grateful to any observers into whose hands the List may come, if they will kindly draw his attention to any such defects, so that they may be remedied in subsequent editions.

The Classification and Nomenclature adopted are very

nearly those of the Fifth Edition of Dr. Gray's Manual of the Botany of the Northern United States, and the writer most gratefully acknowledges the great assistance he has received from the admirable descriptions in that work.

Except in a very general way, no attempt has been made to define the limits of the range of the various Species, as observations tend to show that the range, in many cases, is undergoing constant alteration from various causes. When, however, a Species has appeared to be confined to a particular locality, mention has been made of that fact, but, as a rule, Species known to be of rare occurrence have been excluded.

Characters considered to be of special importance in the determination of the various Species have been emphasized by the use of italics, and where the Species of a Genus, or the Genera of an Order, are numerous, a system of grouping according to some prominent character has been adopted, so as to reduce the labour of determination as much as possible.

To assist the non-classical student, names which might be mispronounced have been divided and accentuated, the division having no reference whatever to the etymology of the words, but being simply based upon their sound when properly pronounced.

It need hardly be added that the writer's ELEMENTS OF STRUCTURAL BOTANY is designed to be the constant companion of the present Flora, in the hands of the young student, for the explanation of such technicalities as he may not have previously mastered.

BARPIE, November, 1883.

Assuming that the student has carefully read the Introductory part of this work, and is familiar with the ordinary botanical terms, and the chief variations in plant-structure as there set forth, it should, with the aid of the accompanying Key, be a very simple task to refer to its proper Family any Canadian wild plant of common occurrence. To illustrate the method of using this Key, let us suppose that specimens of the following plants have been gathered, and that it is desired to ascertain their botanical names, that is, the name of the Genus and the Species of each :—Red Clover, Strawberry, Blue Flag, and Cat-tail Flag.

All of these produce flowers of some kind, and must therefore be looked for under the head of FLOWERING, OR PHANEROGAMOUS PLANTS.

With the specimen of Red Clover in hand, and the book open at page xi., we find that we have first to determine whether our plant is Exogenous or not. The veining of the leaves suggests that it is so; and this impression is confirmed by the fact that the parts of the flower are in fives. Then, is the plant an ANGIOSPERM? As the seed will be found enclosed in an ovary, we answer—Yes. Has the plant both calyx and corolla? Yes. Are the parts of the corolla separate? Here a little doubt may arise; but suppose we answer —Yes. Then our plant will be found somewhere among the POLYPETALOUS EXOGENS. Proceeding with the enquiries suggested under this heading :—Are the stamens more than twice as many as the petals? We find that they are not. Turn then to the heading marked B, page xiii., "Stamens not

more than twice as many as the petals." Under this we find
two subordinate headings, designated by asterisks * and **.
The first of these is not applicable to our plant. Under the
second, marked thus **, we find two minor headings, designated by daggers, ╂ and ╂ ╂. The first of these, "*Corolla irregular*," is clearly the one we want. We have now, therefore,
five Families to select from. We cannot choose any one of
the first four, because our plant has ten stamens, but the characters of the fifth are precisely the characters exhibited by
Clover. Our Clover, therefore, belongs to the Order LEGUMI-
OSÆ. Turning to page 30, and running through the "Synopsis
of the Genera" as there given, we observe that No 2, TRIFO-
LIUM, is the only Genus in which the flowers are in *heads*.
Clover answers the description in the other respects also—viz.:
"leaves of three leaflets," and "stamens diadelphous." The
only question then remaining is, which Species of TRIFOLIUM
have we in hand? Turning to page 31, we find we have
three Species to choose from. No. 2, TRIFOLIUM *pratense*,
is the only one of them with *purplish flowers*, TRIFOLIUM
pratense must, consequently, be the botanical name we are
looking for.

Possibly the observer may decide that the parts of the
corolla are not separate from each other, because in some
instances it is really a doubtful question. He must then
turn to page xvii., and under DIVISION II., GAMOPETALOUS
EXOGENS, he must pursue his inquiries as before. Is the
calyx superior? Plainly not. Proceed then to the heading
B, " Calyx inferior." Are the stamens more than the lobes of
the corolla? Yes. Then the choice of the six Orders in the
Section marked * is easily made as before, and the plant is
referred to LEGUMINOSÆ.

Now let us take the Strawberry. As with Clover we decide without difficulty that the plant is an Exogen. The
carpels are separate, and produce achenes in fruit; the plant
must, therefore, be an ANGIOSPERM. And there is no doubt

that it is Polypetalous.   As the stamens are very numerous it must come under the section marked A.   Under this
section we have three subordinate headings, marked by one,
two, and three asterisks, respectively.   The stamens are
clearly inserted on the calyx, and so our plant must be
found under the heading marked **.   Without hesitation,
we refer it to the Order ROSACEÆ.   Turning to page 34, we
find fourteen Genera to select from.   A very little consideration will show us that No. 8, FRAGARIA, is the Genus we
must fix upon.   Referring to page 39, we have to choose between two species, *Virginiana* and *vesca*, and the choice is
found to depend upon such obvious characters as to furnish
no difficulty.

The leaves of Blue Flag are straight-veined ; the parts of
the flower, also, are in threes.   We therefore decide that the
plant is Endogenous, and on turning to page xxiii., we find
three Divisions of Endogens.   The Flag clearly belongs to
DIVISION II., PETALOIDEOUS ENDOGENS.   Then, is the perianth superior or inferior ?   Clearly the former.   Next, are
the flowers diœcious or perfect ?   Clearly perfect.   And as
the flower has three stamens, it must belong to the Order
IRIDACEÆ, described on page 132.   The Genus is at once seen
to be IRIS, and as only one Species is mentioned, it is probably the one we have in hand—IRIS *versicolor*.

The Cat-tail Flag is also manifestly Endogenous, from the
veining of the leaves.   But it is not a Petaloideous Endogen.
The flowers are collected on a more or less fleshy axis at the
top of a scape.   It therefore belongs to the SPADICEOUS
DIVISION, in which there are four Orders.   The only practical question is, whether our plant belongs to ARACEÆ or
TYPHACEÆ.   On the whole, we choose the latter, and find our
decision confirmed on reading the fuller account of the two
Orders on pages 123 and 124.   The Genus is immediately
seen to be TYPHA, and the Species *latifolia*.

These examples need not be extended here; but the beginner is recommended to run down, in the same manner, a few
plants whose names he already knows. If successful in these
attempts, he will naturally acquire confidence in his deter-
minations of plants previously unknown to him.

KEY

FAMILIES OR ORDERS

INCLUDED IN THIS WORK.

SERIES I. PHANEROGAMS.

Plants producing true flowers and seeds.

CLASS I. DICOTYLEDONS or EXOGENS.

Distinguished ordinarily by having net-veined leaves, and the parts of the flowers in fours or fives, very rarely in sixes. Wood growing in rings, and surrounded by a true bark. Cotyledons of the embryo mostly two.

SUB-CLASS I. ANGIOSPERMS.

Seeds enclosed in an ovary.

DIVISION I. POLYPETALOUS EXOGENS.

Two distinct sets of Floral Envelopes. Parts of the corolla separate from each other.

A. Stamens more **than twice as many as the petals.**

★ Stamens hypogynous (inserted on the receptacle).

+ Pistil apocarpous (carpels separate from each other).

RANUNCULACEÆ.—Herbs. Leaves generally decompound or much dissected. **2**

ANONACEÆ.—Small trees. Leaves entire. Petals 6, in 2 sets. **6**

MAGNOLIACEÆ.—Trees. Leaves truncate. Fruit resembling a cone. **6**

MENISPERMACEÆ.—Woody twiners. Flowers diœcious. Leaves peltate near the edge. **6**

Brasenia, in

NYMPHÆACEÆ.—Aquatic. Leaves oval, peltate ; the petiole attached to the centre. **8**

MALVACEÆ.—Stamens monadelphous. Calyx persistent. Ovaries in a ring. **22**

Podophyllum, in

BERBERIDACEÆ.—Calyx fugacious. Leaves large, peltate, deeply lobed. Fruit a large fleshy berry, 1-celled.   **7**

+ + Pistil syncarpous. (Stigmas, styles, placentæ, or cells, more than one.)

Actæa, in

RANUNCULACEÆ, might be looked for here. Fruit a many-seeded berry. Leaves compound. **2**

NYMPHÆACEÆ.—Aquatics. Leaves floating, large, deeply cordate. **8**

SARRACENIACEÆ.—Bog-plants. Leaves pitcher-shaped.   **9**

PAPAVERACEÆ.—Juice red or yellow. Sepals 2, caducous.   **9**

CAPPARIDACEÆ.—Corolla cruciform, but pod 1-celled. Leaves of 3 leaflets. **14**

* * *Stamens perigynous (inserted on the calyx).*

Portulaca, in

* * * *Stamens epigynous (attached to the ovary).*

Nymphæa, in

B. Stamens not more than twice as many as the petals.

* *Stamens just as many as the petals, and one stamen in front of each petal.*

DIVISION. II. GAMOPETALOUS EXOGENS.

Corolla with the petals united together, in however slight a
degree.

A. Calyx superior (adherent **to the ovary**).

* *Stamens united* **by** *their* **anthers.**

* * *Stamens not united together in any way.*

+ *Stamens inserted on the* **corolla.**

* * * *Stamens just as many as the lobes* **of the** *corolla, inserted* **on** *its tube alternately with* **its lobes.**

+ *Ovaries 2, separate.*

APOCYNACEÆ.—Plants **with milky juice. Anthers converg-**
ing round the stigmas. **but not adherent to them.**
Filaments distinct. 99

ASCLEPIADACEÆ.—Plants **with milky juice. Anthers ad-**
hering to the stigmas. **Filaments** monadelphous.
Flowers **in** umbels. 100

+ + *Ovary 4-lobed around the* **base of the** *style.*

Mentha, in
LABIATÆ.—Stamens 4. Leaves opposite, **aromatic.** .. 89
BORRAGINACEÆ.—Stamens 5. **Leaves alternate.** 93

+ + + *Ovary 1-celled; the* **seeds** *on the walls.*

HYDROPHYLLACEÆ.—Stamens 5, exserted. **Style 2-cleft.**
Leaves **lobed and cut-toothed.** 95

GENTIANACEÆ.—**Leaves entire and opposite; or (in Men-**
yanthes) of **3 leaflets.** 98

+ + + + *Ovary with 2 or more cells.*

AQUIFOLIACEÆ.—Shrubs. **Corolla almost polypetalous.**
Calyx minute. Fruit **a red berry-like drupe.**
Parts of the flower chiefly **in fours or sixes.** .. 80

PLANTAGINACEÆ.—Stamens 4. **Pod 2-celled.** Flowers **in**
a close spike. 80

Verbascum, in
SCROPHULARIACEÆ.—Corolla nearly **regular.** Flowers in a
long terminal spike. Stamens 5; **the** filaments,
or some of them, woolly. 84

POLEMONIACEÆ.—Style 3-cleft. Corolla salver-shaped,
with a long tube. Pod 3-celled, few-seeded; seeds
small. 96

DIVISION III. APETALOUS EXOGENS.

Corolla (and sometimes calyx also) wanting.

A. Flowers not in Catkins.

* *Calyx superior (i.e. adherent to the ovary).*

* * *Both sterile and fertile flowers in catkins, or catkin-like heads.*

SALICACEÆ.—Shrubs or low trees. Ovary 1-celled, many-seeded ; seeds tufted with down at one end. .. 117

PLATANACEÆ.—Large trees. *Stipules sheathing the branchlets.* The flowers in heads. 111

MYRICACEÆ.—Shrubs with resinous-dotted, usually fragrant, leaves. Fertile flowers one under each scale. Nutlets usually coated with waxy grains. 116

BETULACEÆ.—Trees or shrubs. Fertile flowers 2 or 3 under each scale of the catkins. Stigmas 2, long and slender. 116

SUB-CLASS II.   GYMNOSPERMS.

Ovules and seeds naked, on the inner face of an open scale : or, in *Taxus*, without any scale, but surrounded by a ring-like disk which becomes red and berry-like in fruit.

CONIFERÆ.—Trees or shrubs, with resinous juice, and mostly awl-shaped or needle-shaped leaves. Fruit a cone, or occasionally berry-like. 120

CLASS II.   ENDOGENS or MONOCOTYLEDONS.

Distinguished ordinarily by **having** straight-veined leaves (though occasionally net-veined ones), and the parts of the flowers in threes, never in fives. Wood never forming rings, but interspersed in separate bundles throughout the stem. **Cotyledon** only one.

DIVISION I.   SPADICEOUS ENDOGENS.

Flowers collected on **a** spadix, with or without **a spathe or** sheathing bract. Leaves sometimes net-veined.

DIVISION II. PETALOIDEOUS ENDOGENS.

Flowers not collected on a spadix, furnished with a corolla-
like, or occasionally herbaceous, perianth.

A. Perianth superior (adherent to the ovary).

* *Flowers diœcious* or *polygamous, regular.*

* * *Flowers perfect.*

DIVISION III. GLUMACEOUS ENDOGENS.

Flowers without a true perianth, but subtended by thin scales called glumes.

SERIES II. CRYPTOGAMS.

Plants without stamens and pistils, reproducing themselves by spores instead of seeds.

CLASS III. ACROGENS.

Stems containing vascular as well as cellular tissue.

THE COMMONLY OCCURRING

WILD PLANTS OF CANADA,

.

AND MORE ESPECIALLY OF ONTARIO.

SERIES I.

FLOWERING OR PHANEROG'AMOUS PLANTS. .

PLANTS producing Flowers (that is to say, Stamens and Pistils, and usually Floral Envelopes of some kind), and Seeds containing an Embryo.

CLASS I. EXOGENS OR DICOTYLE'DONS.

(See Sections 57–60, Part I., for characters of Class.)

SUBCLASS I. ANGIOSPERMS.

Seeds enclosed in a pericarp.

DIVISION I. POLYPET'ALOUS EXOGENS.

Plants with flowers having both calyx and corolla, the latter consisting of petals entirely separate from each other.

ORDER I. RANUNCULA'CEÆ. (CROWFOOT FAMILY.)

Herbs or woody climbers, with an acrid colourless juice.
Parts of the flower separate from each other. Corolla sometimes wanting. Stamens numerous. Pistil (with one or
two exceptions) apocarpous. Fruit an achene, follicle, or
berry. Leaves exstipulate, with the blades usually dissected, and petioles spreading at the base.

Synopsis of the Genera.

1. **Clematis.** Real petals none or stamen-like. Coloured sepals 4 or more,
va vate in the bud. Fruit an achene, with the long and feathery style
attached. Leaves all opposite. Plant climbing by the bending of the
petioles.

2. **Anemo'ne.** Petals none or stamen-like. Coloured (white) sepals imbricated in the bud. Achenes many, in a head, pointed or tailed, not ribbed.
Stem-leaves opposite or whorled, *forming an involucre remote from the
flower.*

3 **Hepat'ica.** Petals none. Coloured sepals 6-9, whitish or bluish.
Achenes many, not ribbed. Leaves all radical. *An involucre of 3 leaves
close to the flower,* and liable to be mistaken for a calyx.

4. **Thalic'trum.** Petals none. Coloured sepals 4 or more, greenish.
Achenes several, angled or grooved. No involucre. Stem leaves
alternate, decompound. Flowers in panicles or corymbs, mostly
diœcious.

5. **Ranun culus.** Sepals 5, deciduous. Petals generally 5, each with a pit
or little scale on the inside of the claw. Achenes many, in heads, short
pointed. Stem-leaves alternate. Flowers solitary or corymbed, mostly
yellow, rarely white.

6. **Cal'tha.** Petals none. Yellow sepals 5-9. Fruit a many-seeded follicle.
Leaves large, glabrous, heart-shaped or kidney shaped, mostly crenate.
Stem hollow and furrowed.

7. **Cop'tis.** Sepals 5-7, white, deciduous. Petals 5-7, yellow, with slender
claws, and somewhat tubular at the apex. Carpels 3-7, on slender
stalks. Fruit a follicle. Flowers on naked scapes. Leaves radical,
evergreen, divided into three wedge-shaped leaflets, sharply toothed. Root
fibrous, golden yellow.

8. **Aquilegia.** Sepals 5, coloured. Petals 5, *each a long hollow spur.* Carpels 5. Follicles erect, many-seeded. Flowers very showy, terminating
the branches. Leaves decompound.

9. **Actæ'a.** Sepals 4-5, caducous. Petals 4-10, with slender claws. Stamens
many, with long filaments. Fruit a many-seeded berry. Flowers in a
short thick raceme. Leaves decompound, leaflets sharply toothed.

10. **Cimicifuga.** Sepals 4-5 caducous. Petals several, small, two-horned
at the apex. Carpels 1-8, becoming pods. Flowers in long plume-like
racemes.

1. CLEMATIS. VIRGIN'S BOWER.

C. Virginia'na. (COMMON VIRGIN'S BOWER.) A woolly-stemmed climber. Flowers in panicled clusters, often dioecious, white. Leaves of 3 ovate leaflets, which are cut or lobed. Feathery tails of the achenes very conspicuous in the autumn. —Along streams and in swamps.

2. ANEMONE. ANEMONY.

1. A. cylin'drica. (LONG-FRUITED A.) Carpels very numerous, in an oblong woolly head about an inch long. Peduncles long, upright, leafless. Stem-leaves in a whorl, twice or three as many as the peduncles, *long-petioled*. Sepals 5, greenish-white. Plant about 2 feet high, clothed with silky hairs. —Dry woods.

2. A. Pennsylva'nica. (PENNSYLVANIAN A.) Carpels fewer and the head not woolly, but pubescent and spherical. *Stem-leaves sessile*, primary ones 3 in a whorl, but only a pair of smaller ones on each side of the flowering branches. Radical leaves 5-7 parted. Sepals 5, obovate, large and white. Plant hairy, scarcely a foot high. Low meadows.

3. A. nemorosa. (WOOD A. WIND-FLOWER.) Plant not more than 6 inches high, nearly smooth, one-flowered. Stem-leaves in a whorl of 3, long-petioled, 3-5 parted. Sepals 4-7, oval, white, or often purplish on the back.—Moist places.

3. HEPATICA. LIVER-LEAF. HEPATICA.

H. acutil'oba. (SHARP-LOBED H.) Leaves with 3 (sometimes 5) acute lobes, appearing after the flowers. Petioles silky-hairy. —Woods in spring.

4. THALICTRUM. MEADOW-RUE.

1. T. anemonoi'des. (RUE-ANEMONY.) Stem low. Stem-leaves all in a whorl at the top. Roots tuberous. Flowers several in an umbel, by which character this plant is easily distinguished from the Wood Anemony, which it otherwise resembles.—South-westward, in spring.

2. T. dioi'cum. (EARLY M.) Stem smooth, pale and glaucous, 1-2 feet high. Flowers dioecious, in ample panicles, purplish or greenish ; the yellow anthers drooping and very conspicuous. Leaves alternate, decompound ; leaflets with 5–7 rounded lobes. —Woods.

3. **T.** Cornuti. (TALL M.) Stem smooth or nearly so, 2–6 feet high. *Leaves sessile;* leaflets very much like No. 2. Flowers white, in compound panicles ; *anthers not drooping ;* filaments club-shaped.—Low wet meadows, and along streams.

5. RANUNCULUS. CROWFOOT. BUTTERCUP.

1. **R. aqua'tilis.** (WHITE WATER-CROWFOOT.) *Foliage under water,* filiform. *Flowers white,* floating, each petal with a little pit on the inside of the claw.

2. R. multif'idus. (YELLOW WATER-CROWFOOT.) Like No. 1, but larger, and with *yellow flowers.*—Ponds and ditches.

3. R. Flam'mula, var. **reptans.** (CREEPING SPEARWORT.) Stem reclining, rooting at the joints, *only 3–6 inches long.* Leaves linear, entire, remote. Flowers yellow, ½ of an inch broad.—Sandy and gravelly shores of ponds and rivers.

4. **R. aborti'vus.** (SMALL-FLOWERED C.) Petals shorter than the reflexed calyx. Stem erect, *very smooth,* slender. Radical leaves roundish, crenate, petiolate ; stem-leaves 3–5 parted, sessile. Carpels in a globular head, each with a minute curved beak.—Shady hill-sides and wet pastures.

5. **R. scelera'tus.** (CURSED C.) Petals about the same length as the calyx. Stem thick, hollow, smooth. Radical leaves 3-lobed ; stem-leaves 3-parted, uppermost almost sessile. *Head of carpels oblong.*—Wet ditches.

6. **R. recurva'tus.** (HOOKED C.) Petals shorter than the reflexed calyx. *Stem hirsute,* with stiff spreading hairs. Radical and cauline leaves about alike, long-petioled. Head of carpels globular, *each with a long recurved beak.*—Woods.

7. R. **Pennsylva'nicus.** (BRISTLY C.) Petals not longer than the calyx. *Stem hirsute.* Leaves ternately divided, divisions of the leaves stalked, unequally 3-cleft. *Head of carpels oblong, with straight beaks,* and so easily distinguished from No. 6.—Wet places.

8. R. re ____. ____ C.) **Petals much** ____ than the ____. ____ **stems ascending,** *putting* ____ *runners* ____ *the summer.* **Leaves** ____ divisions ____ stalked, petioles hairy. Peduncles ____ furrowed.—Wet places.

9. R. ____. (____ C. ____ BUTTERCUP.) ____ much ____ ____ ____ calyx. **Stem erect,** *from a bulb-like base.* ____ ____ ____ ____, on *furrowed peduncles.*—**Pastures.** ____ rare.

10. R. acris. (T____ C. BUTTERCUP.) **Much taller than** N____ 9 ____ ____ much ____ **than the calyx. Stem upright,** *no bulb at the base.* ____ ____.

6. ____. MARSH-MARIGOLD.

C. palustris. ____ ____MARIGOLD.) **Stem** ____ **a** ____ **high,** ____ ____ ____ **very glabrous.** ____ ____ **yellow,** ____ ____ ____—____ **and wet** meadows; a ____ conspicuous plant in early spring.

7. COPTIS. GOLDTHREAD.

C. trifolia. (GOLDTHREAD ____ GOLDTHREAD.) Low and stemless. ____ 1-flowered, with a single bract above the middle. Petals much smaller than the sepals.—On ____ and ____ stumps in cedar-swamps.

8. AQUILEGIA. COLUMBINE.

A. Canadensis. (WILD COLUMBINE.) Stem branching a foot or more in height, ____ ____ ____ ____ ____ leaflets in threes. Flowers nodding, scarlet outside, yellow within.—Rocky woods and ____.

9. ACTÆA. BANEBERRY.

1. A. spicata. (RED B.) *Raceme short,* broadish **and** ____ being about the ____ ____ ____. ____ red.—Rich woods.

2. A. alba. (WHITE B.) *Raceme long,* thin-based. Pedicels thickened in fruit, ____ ____. ____ ____.—Same localities as No. 1.

C. racemo'sa. (BLACK SNAKEROOT.) Stem 3-6 feet high. Resembling a tall Actæa, but easily distinguished by its plume-like raceme of white flowers.—Along Lake Erie.

ORDER II. MAGNOLIA'CEÆ. (MAGNOLIA FAMILY.)

Trees or shrubs, with alternate entire or lobed (not serrate) leaves. Sepals 3, coloured, deciduous. Petals 6-9, deciduous. Stamens hypogynous, indefinite, separate; anthers adnate. Carpels numerous, in many rows on an elongated receptacle. Fruit resembling a cone.

1. LIRIODEN'DRON. TULIP-TREE.

The only Canadian species is

L. Tulipif'era. A large and stately tree, growing to a great height in many parts of the western peninsula of Ontario. Leaves large, truncate, or with a shallow notch at the end. Flowers large, showy, solitary; petals greenish-yellow, marked with orange. Fruit a dry cone, which, at maturity, separates into dry indehiscent fruits, like samaras.

ORDER III. ANONA'CEÆ. (CUSTARD-APPLE FAMILY.)

Trees or shrubs, with alternate and entire leaves, and solitary, axillary, perfect, hypogynous flowers. Sepals 3. Petals 6, in 2 sets, deciduous. Stamens numerous. Carpels few or many, fleshy in fruit.

1. ASIMINA. NORTH AMERICAN PAPAW.

The only Canadian species is

A. tri'loba. (COMMON PAPAW.) Found only in the Niagara Peninsula. A small tree, not unlike a young beech in appearance, and forming thickets near Queenston Heights. Flowers purple, appearing before the leaves; the 3 outer petals much larger than the 3 inner ones. Fruit 2 to 3 inches long, edible.

ORDER IV. MENISPERMA'CEÆ. (MOONSEED FAMILY.)

Woody twiners, with peltate alternate leaves and small diœcious flowers. Sepals and petals yellowish-white, usually

6 of each. th... in front of the sepals. Stamens nu-
... Fruit a drupe, in appearance something like a small
grape, with moon-shaped seeds.

1. MENISPER'MUM. Moonseed.

The only Canadian species is

M. Canadense. (Canadian Moonseed.) A twining plant,
found, though not abundantly, in low grounds in rich woods. It
may be pretty easily recognized by its usually 7-angled thin
leaves, which are *peltate near the edge.* Fruit bluish-black.

Order V. BERBERIDA'CEÆ. (Barberry Family.)

Herbs (or shrubs), with alternate, petiolate, divided
leaves. Sepals and petals in fours, sixes, or eights except
in the genus Podophyllum), with the petals in front of the
sepals. Stamens (except in Podophyllum as many as the
petals, one before each. **Anthers usually** opening by a valve
at the top. Fruit berry-like.

1. CAULOPHYLLUM. Blue Cohosh.

C. thalictroides. (Blue Cohosh.) Plant 1-2 feet high,
very glaucous and dull purple when young. Flowers yellowish-
green, in a terminal small raceme, appearing in spring before the
decompound leaves are developed. Sepals 6, with 3 little bractlets at their base. Petals 6, thick and somewhat kidney-shaped,
much smaller than the sepals. Stamens 6, one before each petal.
Ovary bursting soon after the flowers, and leaving the 2 drupe-
like seeds naked on their rather thick stalks. Fruit bluish, ¼ of
an inch across.—Rich woods.

2. PODOPHYLLUM. May-Apple. Mandrake.

P. peltatum. Stem about 1 foot high. Flowering stems
with one large 7-9 lobed leaf at the end of it, flat, in the centre;
the flowering ones with two leaves, peltate near the edge, the
flower nodding from the fork. Sepals 6, caducous. Petals 6-9,
large and white. Stamens 12-18. Fruit large, oval, yellowish,
not poisonous.—Found in patches in rich woods. The leaves and
roots are poisonous.

Aquatic herbs with cordate or peltate, usually floating, leaves. Floating flowers on long immersed peduncles. Petals and stamens generally numerous.

Synopsis of the Genera.

1. **Brase'nia.** Sepals and petals each 3 (occasionally 4). Stamens 12-24. Leaves oval, peltate.

2. **Nymphæ'a.** Sepals 4-6. Petals numerous, white, imbricated in many rows, gradually passing into stamens, hypogynous or epigynous. Stamens epigynous. Stigmas united in a flappy head.

3. **Nu'phar.** Sepals 5-6, *yellow*. Petals many, small and stamen-like. Stamens under the ovary.

1. BRASENIA. WATER-SHIELD.

B. peltata. Stems and under surface of the leaves coated with jelly. Leaves oval, 2 inches across, peltate. Flowers small, purplish.—Ponds and slow-flowing streams.

2. NYMPHÆA. WATER-LILY.

1. **N. odora'ta.** (SWEET-SCENTED WATER-LILY.) Leaves orbicular, cleft at the base to the petiole, 5-9 inches wide, often crimson underneath. *Flower very* sweet-scented. Ponds and slow streams.

2. **N. tubero'sa.** (TUBER-BEARING W.) Leaves larger and more prominently ribbed than in No. 1, reniform-orbicular, green on both sides. Flower not at all, or only slightly sweet-scented. Rootstocks producing tubers, which come off spontaneously.— Mostly in slow waters opening into Lake Ontario.

3. NUPHAR. YELLOW POND-LILY.

1. **N. ad'vena.** (COMMON Y. P.) Leaves floating or emersed and erect, thickish, roundish or oblong-cordate. Sepals 6.— Stagnant water.

2. **N. lu'teum.** (SMALL Y. P.) Floating leaves usually not more than 2 inches across, the sinus very narrow or closed. Flowers hardly an inch across. Sepals 5.—Northward, in slow waters.

ORDER VII. **SARRACENIA'CEÆ.** (Pitcher-Plant F.)

Bog-plants, easily distinguished by their pitcher-shaped leaves, all radical.

1. SARRACE'NIA. Side-Saddle Flower.

S. purpu'rea. (Purple S. Huntsman's Cup.) Hollow leaves with a wing on one side, purple-veined, curved, with the hood erect and open. Sepals 5, coloured, with 3 small bractlets at the base. Petals 5, fiddle-shaped, curved over the centre of the flower, deep purple. Ovary 5-celled, globose, the short style expanding above into a 5-angled umbrella, with a hooked stigma at each angle. Flowers on naked scapes, nodding.—Bogs.

ORDER VIII. **PAPAVERA'CEÆ.** (Poppy Family.)

Herbs, with coloured juice and alternate leaves without stipules. Flowers polyandrous, hypogynous. Sepals 2, caducous; petals 4-12. Stamens numerous, anthers introrse. Fruit a 1-celled pod, with numerous seeds.

1. CHELIDO'NIUM. Celandine.

C. majus. Petals 4, deciduous, crumpled in the bud. Juice of the plant yellow. Flower-buds nodding. Flowers small, yellow, in a kind of umbel. Fruit a smooth 1-celled slender pod, from which the 2 valves fall away, leaving the parietal placentas as a slender frame-work, with the seeds attached.—Waste places.

2. SANGUINARIA. Blood-root.

S. Canadensis. Petals 8-12, not crumpled in the bud. Flower-buds not nodding. A stemless plant, with a thick rhizome which emits a red juice when cut, and sends up in early spring a single rounded, 5-7 lobed, thickish leaf, and a 1-flowered scape. Flowers white.—Rich woods.

ORDER IX. **FUMARIA'CEÆ.** (Fumitory Family.)

Smooth herbs, with brittle stems, watery juice, dissected leaves and irregular flowers. Sepals 2, very small. Corolla flattened and closed, of 4 petals, the 2 inner united by their tips over the anthers of the 6 stamens. Stamens in 2 sets of 3 each ; filaments often united ; the middle anther of each set 2-celled, the others 1-celled. Fruit a 1-celled pod.

Synopsis of the Genera.

1. **Adlumia.** Corolla 2-spurred. Petals all permanently united. *Plant climbing.*

2. **Dicentra.** Corolla 2-spurred. Petals slightly united, easily separated. *Not climbing.*

3. **Corydalis.** Corolla 1-spurred. Fruit a slender pod, many-seeded.

1. ADLUMIA. CLIMBING FUMITORY.

A. cirrho'sa. A smooth vine, climbing by the petioles of its decompound leaves. Flowers in axillary pendulous clusters, pale pink.—Low and shady grounds.

2. DICENTRA. DUTCHMAN'S BREECHES.

1. **D. Cucullaria.** (DUTCHMAN'S BREECHES.) Leaves all radical, multifid ; these and the slender scape rising from a bulb-like rhizome of coarse grains. Flowers several in a raceme, whitish, spurs *divergent, elongated, acute, straight.*—Rich woods.

2. **D. Canadensis.** (SQUIRREL CORN.) Underground shoots bearing small yellow tubers, something like grains of corn. Leaves very much as in No. 1. Corolla merely *heart-shaped ; spurs very short and rounded.* Flowers greenish-white, fragrant. —Rich woods.

3. CORYD'ALIS. CORYDALIS.

1. **C. au'rea.** (GOLDEN CORYDALIS.) *Stems low and spreading.* Leaves dissected. *Flowers in simple racemes, golden yellow.* Pods pendulous.—Rocky river-margins and burnt woods.

2. **C. glauca.** (PALE CORYDALIS.) *Stems upright, 1-4 feet* high. *Flowers in compound racemes, purplish tipped with yellow.* Pods erect.—Rocky woods.

ORDER X. CRUCIFERÆ. (CRESS FAMILY.)

Herbs with a pungent watery juice, alternate leaves without stipules, and regular hypogynous flowers in racemes or corymbs. Pedicels without bractlets. Sepals 4, deciduous.

Petals 4, forming a cross-shaped corolla. Stamens 6, 2 of them shorter. Fruit a silique, or silicle. (See Chap. IV., Part I., for dissection of typical flower.) The genera are distinguished by the pods and seeds, the flowers in all cases being much alike. The seeds are exalbuminous, consisting entirely of the embryo, which is folded up in a variety of ways. The radicle may be bent so as to lie against the edge of the cotyledons, and the seed when cut through crosswise shows this section ⊝ ; the cotyledons are then said to be accumbent. Or the radicle may be folded against the *back* of the cotyledon, showing this cross-section ⊜ , in which case the cotyledons are said to be *incumbent* ; and if, besides being incumbent, the cotyledons are doubled round the radicle thus ⊚, they are then *conduplicate*.

Synopsis of the Genera.

* Pod a silique (much longer than broad).

1. **Nasturtium.** Flowers white or yellow. Pod terete, oblong-linear or ellipsoid. Seeds in 2 rows in each cell, globular, without a wing. Cotyledons accumbent.

2. **Dentaria.** Flowers white or pale purple. Pod lanceolate, flat, seeds wingless, on broad seed-stalks. *Stem-leaves 1 or 2 in a whorl,* stem naked below. Rootstock toothed or tuberous. Cotyledons accumbent.

3. **Cardamine.** Flowers white or rose-coloured. Pod linear or lanceolate, flat, seeds wingless, on slender seed-stalks. Stem leafy below. Cotyledons accumbent.

4. **Arabis.** Flowers white or whitish. Pod linear or elongated, flattened, the valves mostly with a distinct mid-rib. Stem leafy. Cotyledons accumbent.

5. **Erysimum.** Flowers yellow. Pod linear, distinctly 4-sided. *Pedicels of the pods diverging from the stem.* Leaves simple. Cotyledons incumbent.

6. **Sisymbrium.** Flowers yellow. Pods awl-shaped, or 4-6 sided, close pressed to the stem, the valves 1-3 nerved. *Pods sessile or nearly so.* Leaves runcinate. Cotyledons incumbent.

7. **Brassica.** Flowers yellow. Pod linear or oblong, nearly terete, or 4-sided, with a distinct beak extending beyond the end of the valves. Cotyledons conduplicate.

* * Pod a silicle (comparatively short).

← Silicle compressed parallel with the broad partition, or globular.

8. **Draba.** Flowers white. Pod flat, *twisted when ripe, many-seeded.* Cotyledons accumbent.

9. **Camelina.** Flowers yellow. Pod pear shaped, pointed, valves 1-nerved. Cotyledons incumbent.

＊ ＋ *Silicle compressed contrary to the narrow partition.*

10. **Capsella.** Flowers white. Pod obcordate-triangular, valves boat-shaped, *wingless.* Seeds numerous. Cotyledons incumbent.

11. **Lepidium.** Flowers white or whitish. Pod roundish, very flat, the valves boat-shaped and *winged.* Seeds solitary.

＊ ＋ — *Silicle fleshy, jointed.*

12. **Cakile.** Flowers purplish. Pod 2-jointed, fleshy. Leaves fleshy. Cotyledons accumbent.

1. NASTURTIUM. WATER-CRESS.

1. **N. officina'le.** (WATER-CRESS.) Flowers white. Stem spreading and rooting. Leaves pinnate : leaflets 3-11, roundish or oblong, nearly entire. Pods oblong-linear.—Ditches and streamlets.

2. **N. palustre.** (MARSH CRESS.) Flowers yellow. Stem erect. Leaves pinnately parted, the lobes cut-toothed. Pods ovoid.—Wet places.

2. DENTARIA. TOOTHWORT. PEPPER-ROOT.

1. **D. diphyl'la.** (TWO-LEAVED T.) Flowers white. *Stem-leaves 2, opposite,* ternately divided. Rootstock toothed, pleasantly pungent to the taste.—Rich woods.

2. **D. lacinia'ta.** (LACINIATE T.) Flowers purplish. *Stem-leaves 3 in a whorl.* Rootstock jointed, scarcely toothed.—Along streams.

3. CARDAMINE. BITTER CRESS.

1. **C. rhomboi'dea.** (SPRING CRESS.) Flowers white or (in var. **purpurea**) rose-purple. *Stem tuberous at the base.* Lower leaves round-cordate ; upper nearly lanceolate ; all somewhat angled or toothed.—Wet meadows.

2. **C. pratensis.** (CUCKOO-FLOWER. LADIES' SMOCK.) Flowers white or rose-colour, showy. Stem from a short rootstock. Leaves pinnate, leaflets 7-15, those of the lower leaves rounded and stalked, entire or nearly so.—Bogs.

3. **C. hirsu'ta.** (SMALL BITTER CRESS.) Flowers white, small. Root fibrous. Leaves pinnate, leaflets 5-11, the terminal leaflets largest. Pods erect, slender.—Wet places.

4. AR'ABIS. ROCK CRESS.

1. **A. lyra'ta.** (Low R.) Flowers white, twice as long as the calyx. Radical leaves clustered, pinnatifid, the terminal lobe largest; stem-leaves scattered. Pods slender, erect, spreading. —Rocky or sandy shores.

2. A. **hirsu'ta.** (HAIRY R.) Flowers greenish-white, small, slightly longer than the calyx. Stem-leaves many, rough, sagittate. *Pods erect, straight.* Stems 1-2 feet high, 2 or 3 from the same root.—Rocky shores and dry plains.

3. **A. læviga'ta.** (SMOOTH R.) Flowers white, rather small. Leaves linear or lanceolate, entire or slightly toothed, sagittate, clasping. Pods long and narrow, *recurved-spreading.* Stem glaucous, 1-2 feet high.—Dry hill-sides. Easily recognised by the pods.

4. **A. Canadensis.** (SICKLE-POD.) Flowers whitish, with linear petals, about twice the length of the calyx. Stem-leaves pointed at both ends, downy. Pods 2-3 inches long, *scythe-shaped, hanging.* Stem 2-3 feet high. A striking plant when the pods are fully formed.—Dry woods and ravines.

5. ERYSIMUM. TREACLE MUSTARD.

E. cheiranthoides. (WORM-SEED MUSTARD.) **Flowers** yellow; inconspicuous. Leaves lanceolate, **scarcely toothed,** roughish with appressed pubescence.—Waste wet **places.**

6. SISYMBRIUM. HEDGE MUSTARD.

S. officinale. (HEDGE MUSTARD.) Flowers yellow, small. **Leaves** runcinate. Stem 1-2 feet high, with spreading **branches.** —**A very** common roadside **weed.**

7. BRASSICA. CABBAGE, MUSTARD, Etc.

1. **B.** sinapistrum. (CHARLOCK.) Flowers bright yellow. Stem 1-2 feet high, branching; it and the leaves hairy.—Too common in our grain **fields.**

2. B. nigra. (BLACK MUSTARD.) Flowers sulphur-yellow. Stem 3-6 feet high, **round,** smooth, and branching. Lower leaves lyrate.—Fields and **waste** places.

8. DRABA. Whitlow-Grass.

D. arab'isans. Flowers white. Stem leafy, erectly branched, pubescent. Leaves lanceolate or linear, minutely dentate. Raceme short, erect. Pods half an inch long, twisted when ripe.—Rocky places.

9. CAMELINA. False Flax.

1. **C. sati'va.** (Common F.) Flowers yellowish. Stem 1-2 feet high, straight, erect, branching. Leaves lanceolate, sagittate. Pods pear-shaped, large, margined.—In flax fields.

10. CAPSEL'LA. Shepherd's Purse.

C. Bursa-pasto'ris. Flowers small, white. Root-leaves clustered, pinnatifid ; stem-leaves clasping, sagittate.—A very common weed.

11. LEPID'IUM. Peppergrass.

1. **L. Virgin'icum.** (Wild P.) Flowers small ; *petals present*, white. Stem 1-2 feet high. Leaves lanceolate, the upper linear or lanceolate and entire, the lower toothed or pinnatifid, tapering towards the base. Pods marginless or nearly so, oval or orbicular.—Along railways and roadsides.

2. **L. intermedium.** Distinguished from No. 1 by having the cotyledons incumbent instead of accumbent, and the pods minutely winged at the top.—Dry sandy fields.

3. **L. rudera'le.** *Petals always absent.* More branched than the preceding.

12. CAKI'LE. Sea-Rocket.

C. America'na. (American S.) Flowers purplish. Leaves obovate, fleshy, wavy-toothed. Pod fleshy, 2-jointed.—Seashore, and borders of the Great Lakes.

Order XI. CAPPARIDA'CEÆ. (Caper Family.)

Herbs in Canada), with an acrid watery juice, and alternate palmately-compound leaves. Flowers cruciform. Stamens 8 or more. Pod like that of a crucifer, but only 1-celled.

1. POLANIS'IA. POLANISIA.

The only species in Canada is

P. grave'olens. A strong-scented herb, with a viscid, hairy stem. Leaflets 3. Flowers in terminal racemes. Sepals 4. Petals 4, yellowish-white, narrowed below into long claws. Stamens 8-12, exserted. Pod glandular-pubescent, 2 inches long, linear. —Shore of Lake Ontario, Hamilton to Niagara.

ORDER XII. VIOLA'CEÆ. VIOLET FAMILY.

Herbs, with alternate stipulate leaves. Flowers irregular, the lower of the 5 petals being spurred. Sepals 5, persistent. Stamens 5, the anthers slightly united and surrounding the pistil. Fruit a 1-celled pod, splitting into 3 valves. Seeds in 3 rows on the walls of the ovary. The only genus represented in this country is

VI'OLA. VIOLET.

* *Stemless Violets; leaves and scapes all from rootstocks.*

+ *Flowers white.*

1. **V. blanda.** (SWEET WHITE V.) Lower petal streaked with purple. Leaves round, heart-shaped, or reniform. Petals beardless. Flower sweet-scented.—Swamps and wet meadows, in spring.

+ + *Flowers blue or purple.*

2. **V. cuculla'ta.** (COMMON BLUE VIOLET.) Leaves on very long petioles, cordate or reniform, the sides folded inwards when young. Lateral petals bearded. Spur short and thick.—Low grounds everywhere.

3. **V. sagitta'ta.** (ARROW-LEAVED V.) Smoothish. Leaves cordate, halberd-shaped, or sagittate, slightly toothed, the first ones on short and margined petioles. Side petals bearded.—Dry hill-sides and old pastures.

• • *Leafy-stemmed Violets.*

+ *Flowers yellow.*

4. **V. pubes'cens.** (DOWNY YELLOW V.) Plant downy. Leaves broadly cordate, coarsely serrate ; stipules large, dentate. Lower petals veined with purple. Spur very short.—Rich woods.

5. **V. Canadensis.** (CANADA VIOLET.) Tall, often a foot high. Leaves large, cordate, serrate-pointed. Petals white inside, purplish outside. Spur very short.—Flowering all summer.

6. **V. cani'na, var. sylvestris.** (DOG V.) Low, spreading by runners. Leaves broadly cordate or reniform, *with fringed-toothed stipules.* Spur cylindrical, half as long as the petals, which are *pale purple.*—Wet places.

7. **V. rostra'ta.** (LONG-SPURRED V.) Distinguished at once by its extremely long straight spur. Petals violet-coloured.

ORDER XIII.   **CISTACEÆ.**   (ROCK-ROSE FAMILY.)

Herbs or low shrubs, with simple entire leaves and regular polyandrous flowers. Calyx persistent, usually of 3 large and 2 smaller sepals. Petals 5 or 3, convolute in the bud. Stamens 3–20. Pod 1-celled, 3-valved. Seeds on 3 parietal projections.

Synopsis of the Genera.

1. **Helian'themum.** Petals 5, fugacious. Style none.
2. **Hudso'nia.** Petals 5, fugacious. Style long and slender.
3. **Lech'ea.** Petals 3, persistent. Style none.

1. HELIAN'THEMUM. ROCK-ROSE.

H. Canad'nse. (FROST-WEED.) Flowers of 2 sorts, some solitary, with large yellow corolla and many stamens, the petals lasting but one day after the flower opens; others small, clustered in the axils of the leaves, and apetalous. Leaves lanceolate, downy beneath.—Sandy places.

2. HUDSO'NIA. HUDSONIA.

H. tomento'sa. (DOWNY H.) *Hoary.* Leaves oval or narrowly oblong, short, close-pressed, or imbricated. Flowers small, *yellow, very numerous.*—A little heath-like shrub, on the shores of the Great Lakes, and the River St. Lawrence.

3. LECHEA. PINWEED.

L. minor. (SMALLER P.) Flowers inconspicuous, purplish, loosely racemose, on distinct pedicels. Stem slender, rough with appressed scattered hairs. Leaves scattered, linear. Pods the size of a pin's head.—Dry soil.

ORDER XIV. DROSERACEÆ. (SUNDEW FAMILY.)

Low glandular-hairy marsh herbs, **with** circinate tufted **radical leaves, and** regular hypogynous **flowers borne on a** naked scape. Sepals, **petals,** and stamens, **5 each ; anthers** turned outwards. Styles 3 or deeply 2-parted. **Pod 1-celled,** 3-valved. The only genus with **us is**

DROSERA. **SUNDEW.**

1. **D.** rotundifolia, (ROUND-LEAVED **SUNDEW.) Flowers** small, white, **in a** 1-sided **raceme. Leaves** orbicular, abruptly **narrowed** into the hairy **petiole, clothed with** reddish **glandular** hairs.—Bogs.

2. **D.** longifolia (LONG-LEAVED **S.) has oblong-spatulate leaves gradually** narrowed into **erect naked petioles.—Bogs ; not common.**

ORDER XV. HYPERICACEÆ. (ST. **JOHN'S WORT F.)**

Herbs or shrubs, with opposite entire dotted leaves and no stipules. Flowers regular, hypogynous, mostly **yellow. Sepals 5,** persistent. Petals 5, deciduous. Stamens **mostly** numerous, and usually 3 or 5 or more bundles. **Styles 3–5, sometimes** united. Pod 1-3-celled. Seeds **numerous.**

Synopsis of the Genera.

1. Hypericum. Petals 5, oblong and somewhat oblique, mostly.

2. Elodes. Petals 5, yellowish, rather reddish-purple.

I. HYPERICUM. ST. JOHN'S WORT.

* Pod 3-celled. Styles 3, separate. Petals with black dots.

1. **H.** perforatum. (COMMON ST. JOHN'S WORT.) Stem much branched, producing runners at the base, slightly 2-edged. Leaves oblong-oblong, with transparent dots, easily observed by holding the leaf up to the light. Petals deep yellow. Flowers in open leafy cymes.—Fields.

2. **H.** corymbosum. (CORYMBED ST. S.) round, but so branching as No. 1. Leaves with both the two spotted dots, oblong, somewhat clasping. Flowers smaller, pale yellow, few red.—Damp woods and wet places generally.

* * *Pod 5-celled. Styles more or less united. Stamens very many, in 5 clusters, if clustered at all.*

3. **H. pyramida'tum.** (GREAT ST. JOHN'S WORT.) Stem 3–5 feet high. Leaves 2–3 inches long, somewhat clasping. *Flowers very large, the petals about an inch long,* and narrowly obovate. Stamens showy. Pod conical, large.—Along streams ; not common.

4. **H. Kalmia'num.** (KALM'S S.) *Shrubby, a* foot or more in height ; leaves linear-lanceolate, crowded, revolute on the margins, thickly placed and sessile. Flowers about 1 inch across, in clusters.—Niagara Falls and westward.

* * * *Pod 1-celled, purple.*

5. **H. ellip'ticum.** (ELLIPTICAL-LEAVED S.) Stem about 1 foot high, *not branched.* Leaves spreading, elliptical-oblong, obtuse, thin. Flowers rather few, showy. in a nearly naked cyme. Pod purple, ovoid, obtuse. Petals pale yellow.—Banks of stream , eastward.

6. **H. muti'lum.** (SMALL S.) Stem slender, branching above, hardly a foot high. Leaves 5-nerved. Cymes leafy at the base. *Flowers* small, not $\frac{1}{4}$ of an inch across.—Low grounds.

7. **H.** Canadense. (CANADA S.) Stem upright, 6–15 inches high, with branches erect. Leaves linear, or linear-lanceolate, 3-nerved at the base, the upper ones acute, sessile. Cymes naked. Pod much longer than the calyx. Flowers small, deep yellow.— Wet sandy places.

8. **H. Saro'thra.** (ORANGE-GRASS.) Stem much branched, 4–9 inches high, branches erect, filiform ; leaves minute, awl-shaped, bract-like. Flowers very small, scattered along the branches, sessile.—Essex County.

2. ELÒDES. MARSH ST. JOHN'S WORT.

E. Virginica. Stem 1–2 feet. Leaves oblong or oval, clasping, marked with pellucid dots, and more or less dotted beneath. Flowers flesh-colored, in clusters in the forks and round of the stem. The stamens in sets of 3, purple-colored. Stamens.

ORDER XVI.— CARYOPHYLLA'CEÆ. (PINK FAMILY.)

Herbs with opposite and entire leaves, *the* . . . . *at the joints.* Flowers . . . . with the parts mostly in . . . . occasionally in fours. . . . . more than twice as many as the petals. . . . 2–5, ... the inner side. . . . usually 1-celled, with the . . . . to the base, or to a column which . . . . from the . . . . of the cell. (Part I., Fig. 15 . .)

Synopsis of the Genera.

• Sepals united into a *tube* or cup. *Petals and stamens borne on the stalk of the ovary; petals with . . . narrow claws.*

1. **Saponaria**. Calyx cylindrical. Styles 2.

2. **Silene**. Calyx toothed. Styles . .

3. **Lychnis**. Calyx 5-toothed. Styles . .

• • *Sepals separate to the base or nearly so. Petals without claws, they and the stamens inserted at the base of the sessile ovary.*

4. **Arenaria**. Petals not cleft at the apex. Styles usually 3. Pod splitting . . . or 6 valves.

5. **Stellaria**. Petals 2-cleft at the apex. Pod splitting to the base into twice as many valves as there are styles. Styles generally 3.

6. **Cerastium**. Petals 2-cleft, or notched. Styles 5. Pod opening at the apex by 10 teeth.

1. SAPONARIA. SOAPWORT.

S. officinalis. (BOUNCING BET.) A stout plant, with rose-coloured or pinkish flowers clustered in corymbs. Leaves 3-ribbed, the lower ovate, upper lanceolate. Pod raised on a short stalk. Styles 2.— Old gardens and roadsides.

2. SILENE. CATCHFLY.

1. **S. inflata.** (BLADDER CAMPION.) . . . . smooth. Stem erect, a foot high. Leaves ovate . . . . (Cap . . . . inflated) ribbed . . . Stamens and styles . . . . —N. . . . . westward.

2. **S. antirrhina** (SLEEPY C.) Stem slender, simple or slightly branching above, a portion of the upper internodes sticky. Leaves linear or lanceolate. Flowers small, pink or purplish, opening only for a short time in sunshine. Calyx ovoid, shining.— Dry soil.

3. **S.** noctiflora. (NIGHT-FLOWERING CATCHFLY.) *Stems* sticky, *pubescent.* Lower leaves spathulate, upper lanceolate. Flowers few, *peduncled.* Calyx-tube with awl-shaped teeth. Petals white or whitish, 2-parted. Opening only at night or in cloudy weather.—A very common weed in cultivated grounds.

4. **S. Virgin'ica** (FIRE. PINK) occurs in South-Western Ontario, and may be recognized by its crimson petals, and bell-shaped calyx, nodding in fruit.

3. LYCHNIS. COCKLE.

L. Githa'go. (CORN COCKLE.) Plant clothed with long soft appressed hairs. *Calyx lobes extremely long,* very much like the upper leaves, surpassing the *purple petals.*—Wheat fields.

4. ARENARIA. SANDWORT.

1. **A. serpyllifo'lia.** (THYME-LEAVED S.) Much branched, 2-6 inches high, roughish-pubescent. Leaves small, ovate, acute. Petals white, hardly as long as the sepals. Sepals pointed, 3-5-nerved. Pod pointed, 6-toothed.—Sandy fields.

2. **A. stricta.** Stems erect, or diffusely spreading from a small root. Leaves awl-sh ped or bristle-form, the upper ones reduced to 1-nerved bracts, crowded in the axils. Cyme diffuse, many-flowered. Sepals pointed, *3-ribbed,* half as long as the white petals.—Rocky fields.

3. **A.** lateriflora. Stem erect, slender, minutely pubescent. Leaves oval or oblong, ½-1 inch long. Peduncles usually three-flowered. Sepals obtuse. Petals white, large, twice as long as the sepals. Flower ¼ of an inch across when fully expanded.—Gravelly shores.

4. **A. peploi'des.** with very fleshy stem and leaves, the latter somewhat clasping, occurs eastward towards the seacoast.

5. STELLARIA. CHICKWEED. STARWORT.

1. **S.** media. (common CHICKWEED.) Stems branching, decumbent, soft and brittle, *marked* lengthwise with one or two pubescent lines. Lower leaves on hairy petioles, acute. Flowers small, white. Petals shorter than the sepals.—Extremely common in damp grounds and old gardens.

2. S. longifo'lia. (LONG-LEAVED STARWORT.) Stems branching, very weak and brittle, supporting themselves on other plants. *Leaves linear.* Pedicels of the flowers long, slender, and spreading, reflexed. Petals white, longer than the 3-nerved sepals.—Low grassy banks of streams.

6. CERASTIUM. MOUSE-EAR CHICKWEED.

1. C. vulga'tum. (COMMON M.) Stem ascending, hairy and somewhat clammy. Leaves ovate or obovate, obtuse. Flowers in close clusters. *Pedicels not longer than the sepals.* Petals shorter than the calyx.—Not common, sometimes confounded with No. 2.

2. C. visco'sum. (LARGER M.) Stems hairy, viscid, spreading. Leaves lanceolate-oblong, rather acute. Flowers in loose cymes. Pedicels longer than the sepals. Petals equalling the calyx.—Fields and copses common.

3. C. arven'se. (FIELD CHICKWEED.) Stem decumbent at the base, pubescent, slender, 4-8 inches high. Leaves linear, or linear-lanceolate, often fascicled in the axils, longer than the lower internodes. Petals obcordate, more than twice as long as the calyx. Pod scarcely longer than the calyx. Cyme few-flowered.

ORDER XVII. PORTULACACEÆ. (PURSLANE F.)

Herbs with fleshy entire exstipulate leaves, and regular hypogynous or perigynous flowers. Sepals 2. Petals 5. Stamens 5-20. Styles 3-8, united below. Pod 1-celled, few- or many-seeded.

Synopsis of the Genera.

1. Portula'ca. Stamens 8-20. Pod opening by a lid (Fig. 161, Part I.), many-seeded.

2. Claytonia. Stamens 5. Pod 2-valved, 3-6-seeded.

1. PORTULACA. PURSLANE.

P. olera'cea. (COMMON PURSLANE.) A low fleshy herb, very smooth, with obovate or wedge-shaped leaves. Calyx 2-cleft, the sepals keeled. Petals yellow, fugacious.—A common pest in gardens.

2. CLAYTONIA. SPRING-BEAUTY.

1. **C. Virgin'ica.** Leaves *linear-lanceolate*, 3–6 inches long.

2. **C. Carolinia'na.** Leaves *ovate-lanceolate* or oblong, tapering at the base. In both species the corolla is rose-coloured, with darker veins. The stem springs from a small tuber, and bears two opposite leaves and a loose raceme of flowers.—Rich woods in early spring.

ORDER XVIII. MALVA'CEÆ (MALLOW FAMILY.)

Herbs, with palmately-veined alternate stipulate leaves. Flowers regular. Calyx valvate. Corolla convolute in the bud. Sepals 5, united at the base. Petals 5, hypogynous. Stamens numerous, monadelphous, hypogynous ; anthers 1-celled. Carpels united in a ring, separating after ripening. Seeds kidney-shaped.

Synopsis of the Genera.

1. **Malva.** Carpels without beaks, 1-seeded. A circle of 3 bractlets at the base of the calyx.

2. **Abutilon.** Carpels 2-beaked, 1–6 seeded. No circle of bractlets.

1. MALVA. MALLOW.

1. **M. rotundifo'lia.** (ROUND-LEAVED MALLOW.) Stems several, procumbent, from a stout tap-root. Leaves long-petioled, round-heart-shaped, crenate, crenately-lobed. Petals obcordate, whitish, streaked with purple, twice as long as the sepals.—Waysides and cultivated grounds.

2. **M. Sylvestris.** (HIGH M.) *Stem erect*, 2 feet high. *Leaves sharply 5–7-lobed. Petals purple*, 3 times as long as the sepals.—Near dwellings.

3. **M. moscha'ta.** (MUSK M.) Stem erect, 1 foot high. *Stem-leaves 5-parted, the divisions cleft.* Flowers large and handsome, rose-coloured or white, on short peduncles, crowded on the stem and branches.—Roadsides near gardens.

2. ABUTILON. INDIAN MALLOW.

A. Avicennæ. (VELVET-LEAF.) Stem 2–5 feet high, branching. Leaves velvety, round-cordate, long-pointed. Corolla yellow.—Near gardens ; not common.

ORDER XIX. TILIA'CEÆ. (LINDEN FAMILY.)

Trees with fibrous bark, soft and white wood, and heart-shaped and serrate leaves, with deciduous stipules. Flowers in small cymes hanging on an axillary peduncle, to which is attached a leaf-like bract. Sepals deciduous. The only Canadian genus is

TIL'IA. BASSWOOD. WHITEWOOD.

T. America'na. (BASSWOOD.) A fine tree, in rich woods. Flowers yellow or cream-coloured, very fragrant. Leaves smooth and green on both sides, obliquely cordate or truncate at the base, sharply serrate. Sepals 5. Petals 5. Fruit a globular nut, 1-celled, 1 2-seeded.

ORDER XX. LINA'CEÆ. (FLAX FAMILY.)

Herbs with entire exstipulate leaves, and regular hypogynous flowers. Sepals petals, stamens, and styles, 5 each. Filaments united at the base. Pod 10-celled, 1 seeded. Our only genus is

LINUM. FLAX.

1. **L. Virginia'num.** (VIRGINIA F.) Flowers yellow, small (¼ of an inch long, scattered. Stem erect, it and the spreading branches terete. Leaves lanceolate and acute, the lower obtuse and opposite.—Dry soil.

2. **L. Stria'tum** has the branches *wing-a g'ed*, broader leaves and more crowded flowers than No. 1. The whole plant is stouter.

3. **L. usitatissimum** (COMMON F.) Flowers blue. Leaves alternate linear-lanceolate acute 3-veined. Cultivated grounds.

ORDER XXI. GERANIA'CEÆ. (GERANIUM FAMILY.)

Strong-scented herbs with pentamerous and symmetrical flowers, the filaments usually united at the base, and 5 glands on the receptacle alternate with the petals. Style 5-cleft. Carpels 5, each 1-seeded, but 1-celled, they, or the lower part of the long styles, attached to a prolonged axis in the receptacle. In fruit the styles often separate from the axis and carry up, or enclose, the carpels with them.

Synopsis **of the** Genera.

1. **Geranium.** Stamens 10, all with anthers.
2. **Ero'dium.** Stamens with anthers only 5.

I. GERANIUM. CRANESBILL

1. **G. macula'tum.** (WILD C.) Stem erect, hairy, about a foot high. Leaves 5–7 parted, the wedge-shaped divisions lobed and cut. Flowers purple, an inch across. *Petals entire, bearded on the claw, much longer than the long-pointed sepals.*—Open woods and fields.

2. **G. Carolinia'num.** (CAROLINA C.) Stem usually decumbent, hairy. Sepals *awn-pointed*, as long as the notched rose-coloured petals.—Waste places.

3. **G. Robertia'num.** (HERB ROBERT.) Stems reddish, spreading, pubescent; branches weak. *Leaves 3-divided*, or *pedately 5-divided*, the divisions twice pinnatifid. Sepals awned, shorter than the reddish-purple petals. *Plant with a very strong odour.*—Shaded ravines and moist woods.

2. ERO'DIUM. STORKSBILL.

E. cicuta'rium. Stem low and spreading, hairy. *Leaves pinnate*, the leaflets sessile, pinnatifid. Peduncles several-flowered. Styles when they separate from the beak *bearded on the inside.*—Not common.

ORDER XXII. **OXALIDA'CEÆ.** (WOOD SORREL F.)

Low herbs with an acid juice and alternate compound leaves, the 3 leaflets obcordate and drooping in the evening. Flowers very much the same in structure as in the preceding Order, but the fruit is a 5-celled pod, each cell opening in the middle of the back (loculicidal), and the valves persistent. Styles 5, separate. The only genus is

OXALIS. WOOD SORREL.

1. **O. acetosel'la.** (WHITE WOOD-SORREL.) Scape 1-flowered. Petals *white, with reddish veins.*—Cold woods.

2. **O. stricta.** (YELLOW W.) Peduncles 2–6-flowered, longer than the leaves. *Petals yellow.* Pod elongated, erect in fruit. —Copses and cultivated grounds.

ORDER XXIII. **BALSAMINA'CEÆ.** (BALSAM FAMILY.)

Smooth herbs, with succulent stems and simple exstipulate leaves. Flowers irregular, the sepals and petals coloured alike, *one of the coloured sepals spurred, the spur with a tail.* Stamens 5, coherent above. Pod bursting elastically and discharging its seeds with considerable force. The only genus is

IMPA'TIENS. TOUCH-ME-NOT. JEWEL-WEED.

1. **I. fulva.** (SPOTTED TOUCH-ME-NOT.) *Flowers orange-coloured, spotted with reddish brown.* Sac longer than broad, conical, tapering into a long recurved spur.—Cedar swamps and along streams.

2. **I. pal'lida.** (PALE T.) Flowers *pale yellow, sparingly dotted with brown.* Sac dilated, broader than long, ending in a short spur.—Wet places.

ORDER XXIV. **RUTA'CEÆ.** (RUE FAMILY.)

Shrubs, with compound transparently-dotted leaves, and an acrid taste. Flowers (with us) diœcious, appearing before the leaves. Stamens hypogynous, as many as the petals. Our only genus is

ZANTHOX'YLUM. PRICKLY ASH.

Z. America'num. (NORTHERN PRICKLY ASH. TOOTH-ACHE TREE.) A prickly shrub with yellowish-green flowers in dense umbels in the axils. Sepals obsolete or none. Petals 5. Stamens in the sterile flowers 5. Carpels 3–5, forming fleshy 1-2-seeded pods. Bark very pungent and aromatic. Leaves pinnate, 4–5 pairs, with an odd one at the end.—Forming thickets in low grounds along streams.

ORDER XXV. ANACARDIA'CEÆ. (CASHEW FAMILY.)

Trees or shrubs, with a milky or resinous juice, and alternate leaves without dots or stipules. Sepals, petals, and stamens, each 5. Fruit a 1-seeded drupelet. The petals and stamens inserted under the edge of a disk which surrounds the base of the ovary. The only genus is

RHUS. SUMACH.

1. **R. typh'ina.** (STAGHORN SUMACH.) A small tree, 10-30 feet high, with *densely so t-hairy branches and stalks.* Flowers greenish-white, polygamous, forming a terminal thyrse. Fruit globular, *covered with crimson hairs.* Leaves pinnate, leaflets 11–31, oblong-lanceolate, serrate, pointed.—Dry hill-sides.

2. **R. glabra** (SMOOTH S.) is *smooth* and seldom exceeds 5 feet in height.

3. **R. Toxicoden'dron.** (POISON IVY.) Shrub about a foot high, smooth. Leaves 3-foliolate, leaflets rhombic-ovate, notched irregularly. Flowers polygamous, in slender axillary panicles. Plant poisonous to the touch.

4. **R. aromat'ica** (FRAGRANT S.) is a shrub 2-3 feet high, with 3-foliolate leaves, sweet-scente l when crushed, and catkin-like spikes of flowers appearing before the leaves — Not common.

ORDER XXVI.   **VITA'CEÆ.**   (VINE FAMILY.)

Shrubs climbing by tendrils, with small greenish flowers in panicled clusters opposite the leaves. Stamens as many as the petals and opposite them. Calyx minute. Petals 4 or 5, hypogynous or perigynous, very deciduous. Fruit a berry, 1–4 seeded. Leaves palmately veined, or compound.

Synopsis of the **Genera.**

1. **Vitis.** Leaves simple, heart-shaped and variously lobed.
2. Ampelopsis. Leaves compound-digitate, of 5 serrate leaflets.

1. VITIS. GRAPE.

1. **V. Labrus'ca.** (NORTHERN FOX-GRAPE.) Leaves and branches woolly. Berries large, dark purple or amber-coloured.—Moist thickets.

2. **V. cordifo'lia.** (FROST GRAPE.) Leaves smooth or nearly so, bright green on both sides, heart-shaped, sharply serrate. Berries small, blue or black.—Banks of streams.

2. AMPELOPSIS. VIRGINIA CREEPER.

A. quinquefo'lia. A common woody vine in low grounds. Leaves digitate of 5 oblong-lanceolate leaflets. Tendrils with sucker-like discs at the end, by which they cling to walls, trunks of trees, &c. Fruit a small black berry.

ORDER XXVII. **RHAMNA'CEÆ.** (BUCKTHORN FAMILY.)

Shrubs with simple stipulate leaves, and small regular perigynous greenish or whitish flowers. Stamens opposite the petals, and with them inserted on the margin of a fleshy disk which lines the calyx-tube. Fruit a berry-like drupe, or a pod.

Synopsis **of the Genera.**

1. **Rhamnus.** Petals minute, *or none.* Drupe berry-like. Calyx and disk free from the ovary.

2 Ceanothus. Petals white, long-clawed, hooded. Fruit dry, dehiscent. Calyx and disk adherent to the base of the ovary.

1. RHAMNUS. BUCKTHORN.

R. alnifo'lius. A low erect shrub, not thorny, with oval acute serrate leaves, and apetalous flowers. Fruit a 3-seeded berry.—Swamps.

2. CEANOTHUS. NEW JERSEY TEA.

C. America'nus. A shrubby plant with downy branches, and ovate, 3-ribbed, serrate leaves. Flowers in white clusters at the summit of the naked flower-branches. Sepals and petals white, the latter hooded, and with slender claws. Pedicels also white.—Dry hill-sides.

ORDER XXVIII. **CELASTRA'CEÆ.** (STAFF-TREE F.)

Shrubs with simple stipulate leaves, alternate or opposite, and small regular flowers, the sepals and petals both imbricated in the bud. Stamens 4-5, alternate with the petals, and inserted on a disk which fills the bottom of the calyx. Pods orange or crimson when ripe.

Synopsis of **the Genera.**

1 Euonymus. Flowers perfect. Seeds 1 or 2, buried at the base, and forming a flat side. Peduncles 4-5-left, leaves opposite. Flowers axillary.

2. Celastrus. Flowers polygamous. Petals and stamens 5. Calyx cupshaped. Leaves alternate. Flowers terminal, raceme.

1. EUONYMUS. SPINDLE-TREE.

1. E. America'nus. (STRAWBERRY BUSH.) A low, rather straggling shrub, with short-petioled or sessile leaves, the latter

ovate or obovate, pointed. Flowers greenish, with the parts generally in fives. Pods rough-warty, *depressed*, crimson when ripe.—Wooded river-banks and low grounds.

2. **E. atropurpu'reus** (BURNING BUSH) occurs in the west of Ontario, and may be distinguished from No. 1 by its greater size (4–8 feet high), its *long-petioled leaves, purplish flowers*, and smooth pods.

2. CELAS'TRUS. STAFF-TREE.

C. scandens. (WAX-WORK. CLIMBING BITTER-SWEET.) A twining smooth shrub, with oblong-ovate, serrate, pointed leaves. Flowers small, greenish, in terminal racemes. Pods orange-coloured. These burst in autumn and display a scarlet pulpy aril, presenting a highly ornamental appearance.—Twining over bushes on river-banks and in thickets.

ORDER XXIX. **SAPINDA'CEÆ.** (SOAPBERRY FAMILY.)

Trees or shrubs, with compound or lobed leaves, and usually unsymmetrical and often irregular flowers. Sepals and petals 4-5, both imbricated in the bud. Stamens 5-10, inserted on a fleshy disk which fills the bottom of the calyx-tube. Ovary 2–3 celled, with 1 or 2 ovules in each cell.

Synopsis of the Genera.

1. **Staphyle'n.** *Flowers perfect.* Lobes of the coloured calyx, the petals, and the stamens, each 5. *Fruit a 3-celled, 3-lobed, inflated pod.* Leaves pinnately compound.

2. **Acer.** *Flowers polygamous.* Leaves simple, variously lobed, opposite. Calyx coloured, usually 5-lobed. Petals none, or as many as the sepals. Stamens 3-12. *Fruit two 1-seeded samaras joined together, at length separating.*

1. STAPHYLE'A. BLADDER-NUT.

S. trifolia. (AMERICAN BLADDER-NUT.) Shrub, 4–6 feet high. Leaflets 3, ovate, pointed. Flowers white, in drooping racemes, at the ends of the branchlets.—Thickets and hill-sides.

2. ACER. MAPLE.

1. **A.** Pennsylva'nicum. (STRIPED MAPLE.) A small tree, 10–20 feet high, with light-green bark striped with dark lines. Leaves 3-lobed at the apex, finely and sharply doubly-serrate, the lobes taper-pointed. Flowers greenish in terminal racemes,

appearing after the leaves.   Samaras large, with divergent wings.—Rich woods.

2. **A. spica′tum.** (MOUNTAIN MAPLE.) A shrub or small tree, 4-8 feet high, growing in clumps in low grounds.   Leaves 3-lobed, coarsely serrate, the lobes taper-pointed.   Flowers greenish, appearing after the leaves, in dense *upright* racemes. Fruit with small widely-diverging wings.

3. A. saccharinum (SUGAR MAPLE.)   A fine tree, with 3-5 lobed leaves, a paler green underneath, *the sinuses rounded, and the lobes sparingly sinuate-toothed.*   Flowers greenish-yellow, *drooping on slender hairy pedicels*, appearing at the same time as the leaves.   Calyx fringed on the margin.—Rich woods.

4. A. dasycar′pum. (WHITE or SILVER M.) Leaves deeply 5-lobed, the sinuses rather acute, silvery-white underneath, the divisions narrow, sharply toothed.   Flowers in erect clusters, greenish-yellow, *appearing much before the leaves; petals none.* Samaras very large, *woolly when young.*—River banks and low grounds.

5. **A. ru′b-um.** (RED M.) Leaves 3-5 lobed, the sinuses acute.   *Flowers* red, appearing much before the leaves. *Petals* (narrowish.)   Samaras smallish, reddish, on drooping pedicels. A smaller tree than No. 4, with reddish twigs, and turning bright crimson in the autumn.—Swamps.

1. **P. verticilla'ta.** Flowers small, greenish-white, in slender spikes. Stems 4-8 inches high, much branched. *Stem-leaves in ar. 4-5 in a whorl*, the upper ones scattered.—Dry soil.

2. **P. Sen'ega.** (SENECA SNAKEROOT.) Flowers greenish-hite, in a solitary cylindrical close spike. Stems several, from hard knotty rootstock, 6-12 inches high. Leaves lanceolate, with rough margins, alternate.—Dry hill-sides and thickets.

3. **P. polyg'ama.** Flowers rose-purple, showy, fringed, in a *many-flowered* raceme. Stems 5-8 inches high, tufted and very leafy, the leaves linear-oblong or oblanceolate. Whitish fertile flowers on underground runners.—Dry soil.

4. **P. paucifo'lia.** (FRINGED P.) Flowers rose-purple, very showy, fringed, *only 1-3 in number*. Stems 1-4 inches high, from long underground runners, which also bear concealed fertile flowers. Leaves ovate, crowded at the top of the stem.— Dry woods.

ORDER XXXI. **LEGUMINOSÆ.** (PULSE FAMILY.)

Herbs, shrubs, or trees, mostly with compound alternate stipulate leaves, and papilionaceous corollas. (For description of a typical flower see Part I., cap. v.) Stamens usually 10, monadelphous, diadelphous, or distinct. Fruit a legume.

Synopsis of the Genera.

1. **Lupi'nus.** *Leaves palmate lyr-shaped, leaflets 7-9.* Flowers in terminal racemes. *Stamens monadelphous.*

2. **Trifo'lium.** Leaves of 3 leaflets. *Flowers in heads.* Stamens diadelphous.

3. **Meli'lotus.** Leaves pinnate, of 3 leaflets. Flowers in axillary spikes. *Pod ovate, or coiled.* Stamens diadelphous.

4. **Trigonella.** Leaves pinnate, of 3 leaflets and stipules toothed. Flowers in slender axillary racemes. Pod wrinkled, 1-2 seeded. Stamens diadelphous.

5. **Robinia.** Trees. Leaves odd-pinnate, often with spines for stipules, and the leaflets with small stipules. Flowers in hanging axillary racemes. Pod margined on one edge. Stamens diadelphous.

6. **Astragalus.** Leaves odd-pinnate, leaflets numerous. Flowers in dense axillary spikes. Corolla long and narrow. Pod turgid, one or both sutures (see Part I., section 13s) *projecting into the cell, thus partially or wholly dividing the cavity.* Stamens diadelphous.

7. Desmo'dium. Leaves pinnate, **of 3 leaflets.** *Calyx 2-lipped.* Flowers purple or purplish, in axillary or terminal racemes. Pod flat, *the lower margin deeply lobed,* *these and* **the** *pod furnished,* roughened with hooked hairs **causing the pods** to adhere to the clothing, &c. Stamens diadelphous.

8. Lespede'za. **Leaves** pinnate, **of** 3 leaflets. *Calyx 5-cleft.* Pod flat, oval or roundish, *occasionally 2-jointed, but* **only 1-seeded.** Flowers sometimes polygamous. Stamens diadelphous.

9. Vicia. Leaves abruptly pinnate, *the footstalk* **prolonged** *into a tendril.* Flowers axillary. *Style slender, hairy at* **the apex.** Pod 2 several-seeded. Stamens diadelphous.

10. Lath'yrus. Leaves as in Vicia, *Style flat-ish,* **flattened above, and** *hairy down the side opposite the germen.* Stamens **diadelphous.**

11. A'pios. *A twining herb.* *Leaves pinnate, of 5-7* **leaflets.** Keel of the flower slender and coiled inward. Flowers **in dense racemes. Stamens** diadelphous.

12. Amphicarpæ'a. *A low and slender twiner,* the stem clothed **with** brownish hairs. *Leaves pinnate, of 3 leaflets.* Flowers polygamous, those of **the** upper racemes perfect, those near the **base fertile, with the** corolla inconspicuous or **none.** Stamens diadelphous.

13. Baptis'ia. Leaves palmate, of 3 leaflets. *Stamens all separate.* The *keel-petals nearly separate.* Racemes terminating **the bushy branches.**

1. LUPI'NUS. LUPINE.

L. peren'nis. (WILD LUPINE.) Stem **erect, somewhat hairy.** Leaflets 7-9, oblanceolate. Calyx deeply **2-lipped. Pods hairy.** —Sandy soil.

2. TRIFO'LIUM. CLOVER. TREFOIL.

1. T. arvense. (RABBIT-FOOT OR STONE **CLOVER.**) Stem erect, 4-12 inches high, branching. Heads **of whitish** flowers oblong, *very silky and soft.* Calyx-teeth fringed **with** long silky hairs.—Dry fields.

2. T. pratense. (RED CLOVER.) **Stems** and leaves somewhat hairy, the latter marked with **a** pale spot on the upper side. Flowers purplish, in dense heads.—Pastures.

3. T. repens. (WHITE CLOVER.) Smooth, creeping. Heads **of** white flowers rather loose.—Fields everywhere.

3. MEDICA'GO. MEDICK.

1. M. lupuli'na. (BLACK MEDICK.) **Stem** procumbent, downy. Leaflets obovate, toothed at the apex. Flowers yellow. Pods kidney-shaped.—Waste places.

2. **M. sati'va** (LUCERNE) has *purple flowers* in a long raceme, and *spirally-twisted pods.*—Cultivated fields.

4. MELILO'TUS. SWEET CLOVER.

1. **M. officina'lis.** (YELLOW MELILOT.) Stem erect, **2–4 feet** high. Leaflets obovate-oblong. Flowers yellow. Pod drooping, 2-seeded.—Waste places.

2. **M. alba** (WHITE M.) is much like No. 1, but has white flowers.—Escaped from gardens.

5. ROBIN'IA. LOCUST-TREE.

1. **R. Pseudaca'cia.** (COMMON LOCUST.) *Racemes slender, loose.* Flowers white, fragrant. A large tree.

2. **R. visco'sa.** (CLAMMY L.) *Racemes crowded.* Flowers white with a reddish tinge. *Branchlets and leafstalks clammy.* Smaller than No. 1.

6. ASTRAG'ALUS. MILK-VETCH.

1. **A. Canadensis.** (CANADIAN MILK-VETCH.) Stem erect, 1–4 feet high, somewhat pubescent. Leaflets *10 or more pairs,* with an odd one at the end. *Flowers greenish yellow, very numerous.*—River-banks.

2. **A. Coop'eri** has *fewer leaflets,* and *white flowers* in a short spike.—Not common.

7. DESMO'DIUM. TICK-TREFOIL.

1. **D. nudiflo'rum.** Stem smooth. 4–8 inches high. Leaves crowded at the summit of st. rile stems. Flowers in a terminal raceme or panicle, *on a scape which rises from the root.* **Leaflets** broadly ovate.

2. **D.** acumina'**tum.** Stem pubescent. Leaves all crowded at the summit of the stem, *from which the raceme or panicle arises.* Leaflets con picuously pointed.—Rich woods.

3. **D. cuspida'tum.** Stems tall, erect, very smooth. Leaflets ovate-lanceolate, taper-pointed, very large, green on both sides. Flowers and bracts large. Pod 4–6 jointed.—Thickets.

4. **D. panicula'tum.** Stem slender, nearly smooth. Leaflets oblong-lanceolate, tapering to a blunt point. Flowers medium-sized. Pod 3–5-jointed, the joints triangular. Racemes panicled. —Rich woods.

5. **D. Canadense.** Stem erect, *hairy*, tall, furrowed. Leaflets oblong-lanceolate, *with nearly straightish veins*. Flowers large, about ½ an inch long.—*Dry woods.*

8. LESPEDE'ZA. Bush-Clover.

1. **L. hirta.** Stem erect, wand-like, tall, pubescent. Leaflets roundish or oval, *pubescent*. Spikes dense, on *peduncles longer than the leaves*. Corolla yellowish-white, with a purple spot on the standard.

2. **L. capita'ta.** *Peduncles and petioles short.* Leaflets varying from oblong to linear, silky underneath. Flowers in *dense heads ;* corolla as in No. 1. Calyx much longer than the pod.— Both species are found in dry soil.

9. VICIA. Vetch.

1. **V. sati'va.** (Common Vetch or Tare.) Stem simple, somewhat pubescent. Leaflets 10–14, varying from obovate-oblong to linear. *Flowers purple, large, one or two together, sessile in the axils, or nearly so.*—Cultivated fields and waste grounds.

2. **V. Cracca.** (Tufted V.) Downy-pubescent. *Leaflets 2 '–24, oblong-lanceolate, strongly mucronate, Peduncles long, bearing a dense one-sided racemo of blue flowers,* bent downward in the spike, and turning purple before withering.—Borders of thickets and pastures. Chiefly eastward.

3. V. Carolinia'na. Smooth. Leaflets 8–12, oblong. *Peduncles bearing a rather loose raceme of whitish flowers, the keel tipped with blue.*—Low grounds and river-banks.

4. V. America'na. Smooth. *Leaflets 10–14, oval or oval-oblong, very veiny. Peduncles 4–8-flowered, flowers purple.* Moist places.

5. **V. hirsu'ta.** Stem weak. Leaflets 8–16, linear. Peduncles 3–6-flowered. *Peduncles slender.* Chiefly eastward.

10. LATHYRUS. Everlasting Pea.

1. L. maritimus. (Beach Pea.) Stem stout, about a foot high. Leaflets 8–16, oval to obovate. *Stipules broad, halbert-shaped, about as large as the leaflets.* Flowers large, purple.—Sea-coast, and shores of the Great Lakes.

2 **L. veno'sus.** (VEINY E.) Stem 2-3 feet high. Leaflets 10-14. Stipules very small, slender, half arrow-shaped. Flowers numerous. Shady banks, chiefly westward and southward.

3. **L. ochroleu'cus.** (PALE E.) Stem slender. Leaflets 6-8, smooth and glaucous. *Stipules half heart-shaped, large. Corolla yellowish-white.*

4. **L. palus'tris.** (MARSH E.) Stem slender, wing-margined. Leaflets 4-8, lanceolate, linear, or narrowly oblong, sharply mucronate. Stipules small, half arrow-shaped. Corolla blue-purple.—Moist places. Var. **myrtifolius** has oblong-lanceolate leaflets, and pale-purple flowers. Upper stipules much larger than the lower ones.

11. A'PIOS. GROUND-NUT. WILD BEAN.

A. tubero'sa. Flowers brown-purple.—A common twining plant in low grounds.

12. AMPHICARPÆ'A. HOG PEA-NUT.

A. monoi'ca. Flowers white or purplish.—Moist thickets and river-banks.

13. BAPTIS'IA. FALSE INDIGO.

B. tincto'ria. (WILD INDIGO.) Smooth and slender, 2-3 feet high, branching. Leaves nearly sessile. Leaflets wedge-obovate, turning black on drying. Flowers yellow.—Dry soil, Lake Erie coast.

ORDER XXXII. ROSA'CEÆ. (ROSE FAMILY.)

Herbs, shrubs, or trees, with alternate stipulate leaves, and regular flowers. The petals mostly 5 and stamens (mostly more than 10) inserted on the edge of a disk which lines the calyx-tube. (See Part I., sections 33 to 45, for typical flowers.)

Synopsis of the Genera.

SUBORDER AMYGDALEÆ.

1. **Prunus.** Calyx bell-shaped, free from the ovary, deciduous. Fruit a drupe.

SUBORDER ROSACEÆ.

2. **Spirœ'a.** Carpels mostly 5, forming follicles in fruit. Calyx 5-cleft, short. Petals obovate, similar.

3. **Gille'nia.** Carpels and fruit as in Spiræa. Calyx elongated, 5-toothed. Petals slender, dissimilar.

4. **Agrimo'nin.** Carpels 2, forming achenes enclosed in the hardened calyx-tube. Calyx armed with hooked bristles. Flowers yellow, in slender spikes.

5. Geum. Carpels numerous, one-ovuled, becoming dry achenes, the persistent styles becoming to be plumose or naked, and straight or jointed. Calyx 5-parted, bractlets.

6. **Waldsteinia.** Carpels 2-6, forming achenes. In the radicle of 3 wedge-form leaflets. Lobes of the calyx minute and deciduous. Flowers yellow, on leafless stems.

7. **Potentil'la.** Carpels numerous, dry, forming achenes heaped on a dry receptacle, the styles not terminal. Lobes of the calyx with 5 alternating bracts.

8. **Fraga'ria.** Flower as in Potentilla, but receptacle becoming fleshy or pulpy and scarlet in fruit. (See Part I, section 151.) Leaves all radical, of 3 leaflets. Low plants producing runners.

9. Dalibarda. Carpels 10, each 2-ovuled, forming nearly dry drupelets. Calyx 5-6 parted, the divisions lacerate on the others and toothed. Calyx with or lacerate enclosing the fruit. Leaves simple, round heart-shaped. Flowers white, on radical.

10. Rubus. Carpels numerous, 2-ovuled, forming drupelets heaped on the receptacle. (See Part I., section 150. Fruit edible. Calyx without bracts.

11. **Ro-n.** Carpels numerous, bristled, forming achenes enclosed in the fleshy calyx-tube. (See Part I., section 44.)

SUBORDER POMEÆ.

12. Crataegus. Calyx 5-toothed, including (?) becoming fleshy in fruit, enclosing pulpy and one-seeded with two stones. Trees or shrubs, branched. Flowers conspicuously white.

13. Pyrus. Fruit a pome or berry-like, the others several cells of a papery or cartilaginous texture. Pith. Leaves simple. Flowers white, pink-tinged, white or rose-colour.

14. Amelanchier. Fruit 5-celled, becoming dry, with ten as berry reddish-blue. Petals oblong-obovate, white. Shrubs or small trees. Flowers.

I. PIRUS. PEAR. CRAB.

1. **P. America'na.** (Wild Plum.) A thorny tree 8-20 feet high, with orange or red drupes half an inch or more in diameter; and ovate, conspicuously pointed, serrate, veiny leaves. Flowers

white, appearing before the leaves, in umbel-like lateral clusters. Woods and river-banks.

2. **P. pu'mila.** (DWARF CHERRY.) A small trailing shrub, is inches high. Leaves obovate-lanceolate, tapering to the base, toothed near the apex, pale beneath. Flowers in umbels of 2-4, appearing with the leaves. Fruit ovoid, dark red, as large as a good-sized pea.—Sandy or gravelly soil, along the Great Lakes.

3. **P. Pennsylva'nica.** (WILD RED CHERRY.) A tree 20-30 feet high, or shrubby. Leaves oblong-lanceolate, sharply serrate, green both sides. Flowers (appearing with the leaves) in large clusters. the pedicels elongated. Fruit globular, as large as a red currant, very sour.—Rocky thickets, and in old windfalls.

4. **P. Virginia'na.** (CHOKE-CHERRY.) A good-sized shrub, 5-10 feet high. Leaves oval, oblong, or obovate, finely and sharply serrate, abruptly pointed. Flowers in short erect racemes, appearing after the leaves. Fruit red, becoming darker, very astringent.—Woods and thickets.

5. **P. sero'tina.** (WILD BLACK CHERRY.) A large tree, with reddish brown branches. Leaves smooth, varying from oval to ovate-lanceolate, taper-pointed, serrate, with short and blunt incurved teeth, shining above. Flowers in long racemes. Fruit purplish-black, edible.—Woods and thickets.

2. SPIRÆ'A. MEADOW-SWEET.

1. **S. opulifo'lia.** (NINE-BARK.) Shrub 3-7 feet high, the old bark separating in thin layers. Leaves broadly ovate or cordate, 3-lobed, doubly crenate. smooth. Flowers white, in umbel-like corymbs terminating the branches. Follicles 2-5, inflated, purplish.—River-banks.

2. **S. salicifo'lia.** (COMMON MEADOW-SWEET.) Shrub 2-3 feet high, nearly smooth. Leaves wedge-lanceolate, doubly serrate. Flowers white or rose-coloured, in a dense terminal panicle.—Low grounds along streams.

3. **S. tomento'sa** (DOWNY M.) with deep rose-coloured flowers, and the stems and under surface of the leaves densely woolly, occurs eastward towards the sea-coast.

3. GILLENIA. INDIAN PHYSIC.

G. trifolia'ta. (BOWMAN'S ROOT.) Herb with 3-foliolate leaves ; the leaflets ovate-oblong, pointed, rather coarsely serrate ; stipules small, awl-shaped, entire. Flowers white or rose-coloured, in loose few-flowered corymbs.—Rich woods, chiefly southwestward.

4. AGRIMONIA. AGRIMONY.

A. Eupato'ria. (COMMON AGRIMONY.) Stem herbaceous, hairy, 2-3 feet high. Leaves interruptedly-pinnate, larger leaflets 5-7, oblong-obovate, coarsely serrate. Petals yellow, twice as long as the calyx.—Borders of woods.

5. GEUM. AVENS.

1. **G. album.** (WHITE AVENS.) Stem 2 feet high, branching, smoothish or downy. Root-leaves pinnate, the cauline ones 3-divided, lobed, or only toothed. Petals white, as long as the calyx. Achenes bristly, tipped with the hooked lower joint of the style, the upper joint falling away.—Low rich woods and thickets.

2. **G. macrophyllum.** Stem bristly-hairy. Root-leaves interruptedly pinnate, the terminal leaflet very large and round heart-shaped ; cauline leaves with small lateral leaflets and a large roundish terminal one, all unequally toothed. Flowers large ; petals golden-yellow. Receptacle of the fruit nearly naked. Achenes tipped as in No. 1.—Cold rocky woods and low meadows.

3. **G. strictum.** (YELLOW A.) Stem 2-3 feet high, rather hairy. Root-leaves interruptedly pinnate ; stem-leaves 3-5 lobed, leaflets obovate or ovate. Petals yellow, larger than the calyx. Receptacle of the fruit downy. Achenes tipped with the hooked style.—Dry thickets.

4. G. rivale. (WATER or PURPLE AVENS.) Petals purplish-yellow ; calyx brown-purple. Flowers nodding, but the fruiting heads upright. The upper joint of the style feathery, persistent. Stem simple, 2 feet high. Root-leaves lyrate ; stem-leaves few, 3-foliolate, lobed.—Bogs and wet places.

5. G. triflorum. Stem about a foot high, soft-hairy. Flowers 3 or more on long peduncles, purple. Styles not jointed, feathery,

at least 2 inches long in the fruit.—Dry hills and thickets, **not** common.

6. WALDSTEINIA. Barren Strawberry.

W. fragarioïdes. A low plant, 4-6 inches high. Leaflets 3, broadly wedge-form, crenately toothed. Scapes several-flowered. Petals yellow, longer than the calyx.—Dry woods and hill-sides.

7. POTENTILLA. Cinque-Foil.

1. **P. Norve'gica.** (Norway Cinque-Foil.) Stem *erect,* *hairy,* branching above. *Leaves palmate, of 3 leaflets;* leaflets obovate-oblong, coarsely serrate. Flowers in cymose clusters. Petals pale yellow, small, *not longer than the sepals.*—Fields and low grounds.

2. **P. paradox'a,** a plant of spreading or decumbent habit, with *pinnate leaves of 5-9 leaflets,* solitary flowers, small petals, and achenes with an appendage at the base, occurs along the south-western shore of Lake Ontario.

3. **P. Canadensis.** (Canada C.) Stem prostrate or ascending, silky-hairy. *Leaves palmate, of 5 leaflets,* the latter serrate towards the apex. Flowers solitary. Petals yellow, *longer than the sepals.*—Dry soil.

4. **P. argen'tea.** (Silvery C.) Stem ascending, branched at the summit, *whitewoolly.* Leaves palmate, of 5 leaflets, the latter deeply serrate towards the apex, *with revolute margins, and woolly beneath.* Petals yellow, longer than the sepals.—Dry fields and roadsides.

5. **P. argu'ta.** Stem stout, 1-2 feet high, brownish-hairy. Leaves pinnate, of 3-9 oval serrate leaflets, downy underneath. Flowers in dense cymose clusters. Petals yellowish or cream-coloured, deciduous. Plant clammy above.—Dry thickets.

6. **P. anseri'na.** (Silver-Weed.) A low plant, creeping with slender runners. Leaves all radical, interruptedly pinnate; leaflets 9-10 serrate, green above, densely silky beneath. Flowers solitary, on long slender peduncles, bright yellow.—River and lake margins.

7. **P. fruticosa.** (Shrubby C.) Stem erect, shrubby, 1-3 feet high, much branched. Leaves pinnate, of 5-7 leaflets,

closely crowded, *entire*, silky, especially beneath. Flowers numerous, large, yellow, terminating the branches.—Bogs.

8 **P. tridenta′ta** (THREE-TOOTHED C.) is common eastward towards the sea-coast. Stem 4–6 inches high. Leaves rigid, palmate, of 3 wedge-shaped leaflets, *3-toothed at the apex. Petals white.*

9. **P. pa′ustris.** (MARSH FIVE-FINGER.) Stem ascending. Leaves pinnate, of 5–7 lanceolate, crowded, deeply serrate leaflets, whitish beneath. *Calyx an inch broad, dark purple inside. Petals purple.*—Bogs.

8. FRAGA′RIA. STRAWBERRY.

1. **F. Virginia′na.** *Achenes deeply imbedded in pits* on the surface of the fleshy receptacle ; calyx erect after flowering. Leaflets thin.

2. F. ves′ca. *Achenes not sunk in pits,* but merely on the surface of the receptacle ; calyx spreading. Leaflets thin.

9. DALIBAR′DA. DALIBARDA.

D. repens. Stems tufted, downy. Whole plant with something of the aspect of a violet.—Low woods.

10. RUBUS. BRAMBLE.

1. R. odora′tus. (PURPLE-FLOWERING RASPBERRY.) Shrubby, 3–5 feet high. Branches, peduncles, and calyx *clammy with glandular hairs. Flowers large and handsome, rose-purple.* Leaves large, broadly ovate, 3–5 lobed, the lobes acute, minutely toothed. *Fruit flat.*

2. R. triflo′rus. (DWARF RASPBERRY.) Stems ascending or trailing a foot high, not prickly. Leaflets 3–5 nearly smooth, rhombic-ovate, acute at both ends, doubly serrate. Peduncle usually 2-flowered. Petals white ; sepals reflexed. Fruit red.—Cedar swamps.

3. R. strigosus. (WILD RED RASPBERRY.) Stems *upright, beset with stiff straight bristles.* Leaflets 3–5, oblong-ovate, pointed, cut-serrate, whitish beneath. *Fruit light red.*—Hill-sides and thickets.

4. R. occidenta′lis. (BLACK RASPBERRY.) Stems *glaucous, recurved, armed with hooked prickles.* Leaflets 3, ovate, pointed,

coarsely serrate, white-downy beneath. **Fruit** purplish-black.—
Borders of fields, especially where the ground has been burned
over.

5. **R. villo'sus.** (HIGH BLACKBERRY.) Stem shrubby, fur_
rowed, erect or reclining, armed with hooked prickles. **Leaflets**
3–5, unequally serrate, the terminal one conspicuously stalked.
Flowers racemed, numerous, large and white. *Fruit oblong, black.*
—Borders of thickets.

6. **R. his'pidus** (RUNNING SWAMP BLACKBERRY) occurs occasionally in low meadows. Stem prostrate, with small reflexed
prickles, sending up at intervals the short flowering shoots.
Leaflets mostly 3, smooth and shining. Fruit of few grains, red
or purple.

II. ROSA. ROSE.

1. **R. Caroli'na.** (SWAMP ROSE.) Stem 4–8 feet high, erect,
armed with *stout hooked* prickles, but no bristles. Leaflets 5–9,
finely serrate. *Flowers in corymbs, numerous.* *Calyx* and globular
calyx-tube beset with glandular bristles.—Wet places.

2. **R. lu'cida.** (DWARF WILD ROSE.) Stem 1–2 feet high,
armed with slender *almost straight* prickles, and bristles. Leaflets 5–9, finely serrate. *Peduncles 1–3-flowered. Calyx-teeth
bristly, but the tube in fruit nearly smooth.*—Dry soil, or borders
of swamps.

3. **R. blanda.** (EARLY WILD ROSE.) Stem 1–3 feet high.
*Prickles few and scattered, straight. Leaflets 5–7. Peduncles 1–3-
flowered. Calyx and fruit* smooth, the lobes of the calyx erect
and connivent in fruit.—Dry woods and fields.

4. **R. rubigino'sa.** (SWEET-BRIER.) Stem tall. Prickles
numerous, the larger hooked, the smaller awl-shaped. Leaflets
5–7, doubly serrate, glandular beneath. *Flowers mostly solitary.*
Fruit *pear-shaped* or obovate.—Roadsides and fields.

III. CRATÆGUS. HAWTHORN.

1. **C. cocci'nea.** (SCARLET-FRUITED THORN.) **A low tree**,
glabrous. Leaves rather *thin*, roundish-ovate, serrate, on *slender
peduncles.* Fruit bright red, ovoid, hardly edible.—Thickets.

2. **C. tomentosa.** (BLACK or PEAR THORN.) **A tall shrub**
or low tree, downy, at least when young. Leaves *thickish*, oval

or broadly ovate, finely serrate, on *margined petioles*, furrowed along the veins. Fruit globular or pear-shaped, larger than in No. 1, edible.—Thickets.

3. **C. Crus-galli.** (COCKSPUR THORN.) A shrub or low tree, *glabrous*. Leaves thick, shining above, wedge-obovate, finely serrate. *Petioles very short.* Fruit globular, bright red. Thorns very long.—Thickets.

13. PYRUS. PEAR. APPLE.

1. **P. corona'ria.** (AMERICAN CRAB-APPLE.) A small tree, with ovate serrate *simple leaves.* Flowers in umbel-like cymes. Styles woolly and cohering at the base. *Fruit a greenish apple.* —Chiefly west of Toronto.

2. P. arbutifo'lia. (CHOKE-BERRY.) A shrub, with obovate finely serrate *simple leaves.* Flowers in compound cymes. *Fruit berry-like,* nearly globular, dark-red or *black.*—Swamps.

3. **P.** America'na. (AMERICAN MOUNTAIN ASH.) A small tree, with odd-pinnate leaves of 13-15 *leaflets,* the latter lanceolate, taper-pointed, sharply serrate, bright green. *Fruit scarlet, berry-like.* Flowers in flat cymes.—Swamps and cool woods, northward.

14. AMELANCHIER. JUNE BERRY.

A. Canadensis. (SHADBUSH.) A shrub or small tree, with a purplish, berry-like, edible fruit. The variety **Botryapium** has ovate-oblong leaves, very sharply serrate, and white flowers in long drooping racemes, the petals 4 times as long as the calyx. The variety rotundifolia has broader leaves and shorter petals, and 6-10-flowered racemes.

ORDER XXXIII. SAXIFRAGACEÆ. (SAXIFRAGE F.)

Herbs or shrubs, distinguished from Rosaceæ chiefly in having opposite as well as alternate leaves, and usually no stipules; stamens only as many or twice as many as the (usually 5) petals; and the carpels fewer than the petals (mostly 2), and usually more or less united with each other. Stamens and petals generally inserted on the calyx.

1. **Ri'bes.** Shrubs, sometimes prickly, with alternate and palmately-veined and lobed leaves, which are plaited in the bud. Calyx 5-lobed, the tube adherent to the ovary (superior). Petals 5, small, inserted on the calyx. Stamens 5. Styles 2. *Fruit a many-seeded berry.*

2. **Parnassia.** Smooth herbs, with entire and chiefly radical leaves, and solitary flowers terminating the long scapes. Petals 5, large, *veiny, each with a cluster of sterile filaments at the base.* Proper stamens 5. Stigmas 4. Pod 4-valved. Calyx free from the ovary.

3. **Saxifraga.** Herbs with *clustered root-leaves.* Flowers in close cymes. Calyx-lobes hardly adherent to the ovary. Petals 5. Stamens 10. *Fruit a pair of follicles,* slightly united at the base.

4. **Mitella.** Low and slender herbs, with round-heart-shaped radical leaves, *those on the scape (if any) opposite.* Flowers in terminal racemes. Calyx 5-lobed, adherent to the base of the ovary. Petals 5, slender, *pinnatifid.* Stamens 10, *short.* Styles 2. Pod 2-beaked, but 1-celled.

5. **Tiarella.** Slender herbs, with radical heart-shaped leaves, and *leafless scapes,* bearing a simple raceme of flowers. Calyx bell-shaped, 5-parted. Petals 5, *entire.* Stamens 10, *long and slender.* Pod 2-valved, *the valves unequal.*

6. **Chrysosplenium.** Small and smooth herbs, with mostly opposite roundish leaves. Calyx-tube adherent to the ovary. *Petals none.* Stamens twice as many as the calyx-lobes (8-10), inserted on a conspicuous disk. Pod 2-lobed.

1. RI'BES. CURRANT. GOOSEBERRY.

1. **R. Cynos'bati.** (WILD GOOSEBERRY.) Stem with small thorns at the bases of the leaves, the latter downy, on slender petioles, roundish heart-shaped, 3-5 lobed. *Peduncles slender,* 2-3-flowered. *Berry covered with long prickles.*—Open woods and clearings.

2. **R. hirtellum.** (SMALL WILD GOOSEBERRY.) Stems with very short thorns or none. *Peduncles very short,* 1-2-flowered. *Berry small, smooth.*—Low grounds.

3. **R. lacus'tre.** (SWAMP GOOSEBERRY.) Shrubby. Young stems prickly, and thorny at the bases of the leaves. Leaves cordate, *deeply 3-5-lobed, the lobes deeply cut. Racemes 4-9-flowered, slender, nodding.* Fruit bristly.—Swamps and wet woods.

4. **R. flor'idum.** (WILD BLACK CURRANT.) *Stems and fruit without prickles or thorns. Leaves resinous-dotted,* sharply 3-5-lobed, doubly serrate. Racemes many-flowered, drooping. Calyx bell-shaped. *Fruit black,* smooth.—Woods.

5. **R. rubrum.** (WILD RED CURRANT.) A low shrub with straggling stems. Leaves obtusely 3–5-lobed. *Racemes from lateral buds separate from the leaf-buds*, drooping. Calyx flat. *Fruit red*, smooth.—Bogs and wet woods.

2. PARNAS′SIA. GRASS OF PARNASSUS.

P. Carolinia′na. Petals sessile, very veiny. Sterile filaments 3 in each set. Leaves ovate or rounded, *usually only one low down on the stalk.* Flower an inch across.—Beaver meadows and wet banks.

3. SAXIF′RAGA. SAXIFRAGE.

S. Virginiensis. (EARLY SAXIFRAGE.) Stem 4–9 inches high. Scape clammy. Leaves obovate, crenately toothed. Petals white, oblong, twice as long as the sepals.—Damp rocks along streams.

4. MITEL′LA. MITRE-WORT. BISHOP'S CAP.

1. **M. diphyl′la.** (TWO-LEAVED MITRE-WORT.) Stem hairy. Leaves cordate, 3–5-lobed, *those on the scape 2, opposite*, nearly sessile. Flowers white.—Rich woods.

2. **M. nuda.** (NAKED-STALKED M.) Stem small and delicate. Leaves kidney-shaped, *doubly crenate. Scape leafless*, few-flowered. *Flowers greenish.*—Deep woods, on moss-covered logs, &c.

5. TIAREL′LA. FALSE MITRE-WORT.

T. cordifolia. Scape leafless, 5–12 inches high. Leaves heart-shaped, sharply-toothed, sparsely hairy above, downy beneath. Petals white, oblong.—Rich woods.

6. CHRYSOPLE′NIUM. GOLDEN SAXIFRAGE.

C. America′num. A low and delicate smooth herb, with spreading and forking stems. Flowers greenish-yellow, inconspicuous, nearly sessile in the forks.—Shady wet places.

ORDER XXXIV. CRASSULA′CEÆ. (ORPINE FAMILY.)

Succulent herbs (except in our genus), chiefly differing from Saxifragaceæ in having *symmetrical flowers*, the sepals, **petals,** and carpels being the same in number, and the stamens either as many or twice as many. The only genus represented among our common wild plants is

PENTHORUM. DITCH STONE-CROP.

P. sedoi'des. Not succulent. Sepals 5. Petals 5, if any; sometimes wanting. Stamens 10. *Pod 5-angled, 5-horned, and 5-celled.* Leaves scattered, lanceolate, acute at both ends. A homely weed, with greenish-yellow flowers in a loose cyme.— Wet places. (Parts of the flowers occasionally in sixes or sevens.)

ORDER XXXV. HAMAMELA'CEÆ. (WITCH-HAZEL F.)

Tall shrubs, with alternate simple leaves, and deciduous stipules. Flowers in clusters or heads, often monoecious. Calyx 4-parted, adherent to the base of the ovary, the latter of 2 united carpels. Fruit a 2-beaked, 2-celled woody pod, opening at the top. Petals 4, strap-shaped, inserted on the calyx. Stamens 8, 4 of them anther-bearing, the remainder reduced to scales. The only genus with us is

HAMAMELIS. WITCH-HAZEL.

H. Virgin'ica. Leaves obovate or oval, crenate or wavy-toothed, pubescent. Flowers yellow, appearing late in autumn. —Damp woods, chiefly west of Toronto.

ORDER XXXVI. HALORA'GEÆ. (WATER-MILFOIL F.)

Aquatic or marsh plants, with small inconspicuous flowers, sessile in the axils of the leaves or bracts. Calyx-tube adherent to the ovary, the latter (in our genus) 4-lobed, 4-celled. Limb of the calyx minute or none. Petals 4, if any. Stamens 4 or 8. Immersed leaves pinnately dissected into capillary divisions. Plants mostly under water, except the flowers. The only common genus here is

MYRIOPHYL'LUM. WATER-MILFOIL.

1. **M. spica'tum.** *Stamens 8.* Bracts ovate, entire, *shorter than the flowers.* Leaves in whorls of 3 or 4. Flowers greenish, in terminal spikes. Stem very long.—Deep water.

2. **M. verticilla tum.** *Stamens 8.* Leaves finely dissected and whorled as in No. 1. Bracts pectinate-pinnatifid, much longer than the flowers, and the spike therefore leafy. Stem 2-4 feet long.—Stagnant water.

3. **M. heterophyllum.** *Stamens* Lower leaves dissected, in whorls of 4 or 5. Bracts ovate or lanceolate, finely serrate, crowded, the lower ones pinnatifid. Stem stout.—Stagnant or slow water.

ORDER XXXVII. **ONAGRACEÆ.** (EVENING-PRIMROSE F.)

Herbs with perfect and symmetrical flowers, the parts of the latter in twos or fours. Calyx-tube adherent to the ovary and usually prolonged above it. Petals and stamens inserted on the calyx. Style 1. Stigmas 2 or 4 or capitate. (See Part I, sections 39-42, for description of a typical plant.)

Synopsis of the Genera.

1. **Circæa.** *Petals 2, obcordate. Stamens 2.* Stigma capitate. Fruit *bur-like,* 1 beset with hooked bristles. Delicate low plants with opposite and very small white in racemes.

2. **Epilobium.** *Petals 4. Stamens 8.* Calyx-tube hardly prolonged beyond the ovary. Fruit a linear and many-seeded, *the seeds provided with tufts of downy hairs.*

3. **Œnothera.** *Petals 4. Stamens 8.* Stigma 4-lobed. *Flowers yellow.* Calyx-tube much prolonged. Pods cylindrical or club-shaped. *Seeds without tufts.*

4. **Ludwigia.** *Petals 4 or none. Stamens 4.* Calyx-tube not prolonged. Stigma capitate.

1. CIRCÆA. Enchanter's Nightshade.

1. **C. Lutetiana.** Stem 1-2 feet high. Leaves opposite, ovate, slightly toothed. No bracts under the pedicels. Fruit roundish, burr-shaped, 2-celled. Rich woods.

2. **C. Alpina.** Stem *low* and a mere 3-8 inches. Leaves cordate, coarsely toothed. *Minute bracts under the pedicels.* Fruit club-shaped, soft-hairy, 1-celled.—Deep low woods.

2. EPILOBIUM. Willow-herb.

1. **E. angustifolium.** (GREAT WILLOW-HERB.) Stem 3-6 feet high, simple. Leaves lanceolate. Flowers purple, very showy, *in a terminal* **raceme** *or spike.* **Stigma** of 4 long lobes.— Newly-cleared land.

2. E. palustre. Stem 1-2 feet high, erect, slender, branching above, *hoary-pubescent*. Leaves linear, nearly entire. Flowers *small, corymbed* at the ends of the branches, purplish or white. Petals erect. Stigma club-shaped.—Bogs.

3. E. molle is occasionally met with. It differs from No. 2 chiefly in having the leaves crowded and their points more obtuse. The petals are rose-coloured.—Bogs.

4. E. colora'tum. Stem 1-2 feet high, *nearly smooth.* Leaves *lanceolate* or *ovate-lanceolate.* Flowers small, corymbed. Petals *purplish, deeply notched.*—Extremely common in wet places.

3. ŒNOTHE RA. EVENING PRIMROSE.

1. Œ. bien'nis. (COMMON EVENING PRIMROSE.) *Stem 2-4 feet high,* hairy. Leaves ovate-lanceolate. Flowers yellow, *odorous, in a leafy* spike, opening in the evening or in cloudy weather. Pods oblong, narrowing towards the top.—Waste places.

2. Œ. Pu'mila. (SMALL E.) Stem low, *5-12 inches high,* smooth or nearly so. Leaves lanceolate or oblanceolate. Pods nearly sessile, club-shaped, *4-angled.*—River and lake margins.

4. LUDWIG'IA. FALSE LOOSESTRIFE.

L. palustris. (WATER PURSLANE.) Stems creeping in the mud of ditches or river margins, smooth. Leaves opposite, tapering into a slender petiole. Flowers sessile, solitary, usually without petals. Pod 4-sided.

ORDER XXXVIII. **LYTHRA'CEÆ.** (LOOSESTRIFE F.)

Herbs, or slightly woody plants, with opposite or whorled entire leaves, without stipules. Calyx enclosing, but free from, the ovary. Petals (mostly 5) and stamens (mostly 10) inserted on the calyx. Flowers axillary or whorled. Style 1. Stigma capitate. The only common representative genus with us is

NESÆ'A. SWAMP LOOSESTRIFE.

N. verticilla'ta. Stems curving, 2-6 feet long, 4-6-sided. Leaves lanceolate, mostly whorled. Flowers purple, in the axils of the upper **leaves.** Calyx bell-shaped, with 5-7 erect teeth, with **supplementary** projections between them. Stamens **10,** exserted, 5 longer than the rest.—Swamps.

ORDER XXXIX. UMBELLIFERÆ. (PARSLEY FAMILY.)

Herbs with small *flowers* mostly **in compound** *umbels.* Calyx-tube grown fast **to the surface of the ovary ; calyx-teeth** minute **or none.** The 5 petals and 5 stamens **inserted on a disk which crowns the** ovary. Styles 2. **Fruit dry,** 2-seeded. Stems hollow. Leaves **usually** much cut. (See Part I, Chapter VI, for description of a typical flower.)

Synopsis of the Genera.

§ 1. *Seeds flat (not hollowed) on the inner face.*

1. **Hydrocotyle.** *Umbels simple, or one springing from the summit of* another, axillary. Flowers white. Stem slender and creeping. **Leaves** round-kidney-shaped.

2. **Sanicula.** *Umbels irregular (or compound),* the *greenish flowers congregate in the umbellets. Leaves* palmately lobed or parted. Fruit globular, covered with hooked prickles.

(In the Genera which follow, the umbels are regularly compound.)

3. **Daucus.** Stem bristly. Leaves twice or thrice-pinnate, or **pinnatifid.** Bracts of the involucre pinnatifid, very long. Fruit ribbed, the *ribs* bristly.

4. **Heracleum.** *Stem 3-4 feet high, woolly* and grooved. Leaves **1-2-tern-**ately compound. *Flowers white,* the outer **corollas** larger **than the** others. *Fruit wing-margined at* **the** junction **of the** carpels, very flat.

5. **Pastinaca.** *Stem smooth, grooved.* **Leaves pinnate.** *Flowers yellow, all alike.* Fruit as in No. 4.

6. **Archangelica.** Stem smooth, **stout, purple.** Leaves 2-3-ternately compound. *Flowers greenish-white.* Fruit smooth, flattish on the back, double-wing-margined, each carpel with 3 ribs on the back.

7. **Conioselinum.** Stem smooth. Leaves finely 2-3-pinnately compound, the petioles inflated. *Flowers white.* Fruit *doubly wing-margined,* and with 3 narrow wings on the back of each carpel.

8. **Thaspium.** Stem smooth. Leaves 1-2-ternately **divided.** *Flowers deep yellow.* Fruit not flattened, *winged or ribbed.*

9. **Zizia.** Stem slender, smooth and glaucous. Leaves 2-3-ternately com**pound.** *Flowers yellow.* Rays of the umbel long and slender. Fruit contracted at the junction of the carpels, the carpels narrowly 5-ribbed.

10. **Cicu'ta.** Stem streaked with purple, stout. Leaves *thrice compound.* *Flowers white.* Fruit a little contracted at the sides, *the carpels strongly 5-ribbed.*

11. **Sium.** Stem grooved. Leaves simply *pinnate.* *Flowers white.* Fruit as in No. 10.

12. **Cryptotæ'nia.** Stem smooth. *Leaves 3-foliolate. The umbels with very unequal rays.* Flowers white. Fruit nearly as in Nos. 10 and 11.

§ 2. *Inner face of each seed hollowed lengthwise.*

13. **Osmorrhi'za.** Leaves large, 2-3-ternately compound. Flowers white. Fruit linear-oblong, angled, tapering downwards into a stalk-like base. Ribs of the carpels bristly upwards.

§ 3. *Inner face of each seed curved inwards at top and bottom.*

14. **Erige'nia.** Stem low and smooth. Leaves 2-3-ternately divided. *Fruit twin.* Carpels nearly kidney-form. Umbels 3-rayed, small. Flowers white.

1. HYDROCOT'YLE. WATER PENNYWORT.

H. America'na. Stem spreading and creeping, very slender. Leaves kidney-shaped, crenate, slightly lobed. Umbels 3-5-flowered, inconspicuous, in the axils of the leaves.—Shady wet places.

2. SANIC'ULA. SANICLE. BLACK SNAKEROOT.

1. **S. Canadensis.** *Leaves 3-5-parted.* A *few staminate flowers* among the perfect ones, and on *very short pedicels.* Styles shorter than the prickles of the fruit.—Low rich woods, not so common as the next.

2. **S. marilan'dica.** *Leaves 5-7-parted.* Staminate flowers numerous, and *on slender pedicels.* Styles long, recurved.—Rich woods.

3. DAUCUS. CARROT.

D. Caro'ta. (COMMON CARROT.) Found wild occasionally in old fields. In fruit the umbel becomes hollow like a bird's nest.

4. HERACLE'UM. COW PARSNIP.

H. lana'tum. Umbels large and flat. Petioles of the leaves spreading and sheathing. Leaves very large; leaflets broadly heart-shaped, deeply lobed.—Low wet meadows.

5. PASTINA'CA. PARSNIP.

P. sati'va. (COMMON PARSNIP.) Found wild in old fields and along roadsides. Leaflets shining above.

6. ARCHANGELICA. ARCHANGELICA

A. atropurpu rea. (GREAT ANGELICA.) Stem very tall (4 6 feet) and stout, dark purple. Whole plant strong-scented. Petioles much inflated at the base.—Marshes and low river banks.

7. CONIOSELINUM. HEMLOCK-PARSLEY.

C. Canadense. Stem 2-4 feet high. Petioles much inflated. Leaflets of the involucels awl-shaped.—Swamps.

8. THASPIUM. MEADOW-PARSNIP.

T. au'reum. Stem 1-2 feet high, angular-furrowed. Leaves oblong-lanceolate, sharply serrate. Fruit with 10 winged ridges, or in var. apterum with 10 ribs.—Dry or rich woods.

9. ZIZIA. ZIZIA.

Z. integer'rima. Stem slender, 1-2 feet high. Involucels none. Plant strong-scented.—Rocky hill-sides

10. CICUTA. WATER HEMLOCK.

1. C. macula'ta. SPOTTED COWBANE. BEAVER POISON. Stem 3-6 feet high, purplish, smooth. Leaflets coarsely serrate, pointed.—Swamps and low grounds.

2. C. bulbif'era is easily distinguished from No. 1 by bearing clusters of bulblets in the axils of the upper leaves. The leaflets, also, are linear.—Swamps and low grounds.

11. SIUM. WATER PARSNIP.

S. lineare. Stem 2 3 feet high, furrowed. Leaflets varying from linear to oblong, sharply pointed and serrate.—Borders of marshes, usually in the water.

12. CRYPTOTÆNIA. HONEWORT

C. Canadensis. Stem 1 2 feet high, slender. Leaflets large, ovate, doubly serrate. No involucre.—Rich woods and thickets.

13. OSMORRHIZA. SWEET CICELY.

1. O. longis'tylis. (SMOOTHER SWEET CICELY.) Stem reddish, nearly smooth. Leaflets sparingly pubescent, short pointed. Styles slender, nearly as long as the ovary, recurved.—Rich woods

2. **O.** brevistylis. (HAIRY SWEET CICELY.) Whole plant hairy. Leaflets taper-pointed. Styles *very* short, conical.—Rich woods.

14. ERIGENIA. HARBINGER-OF-SPRING.

E. bulbosa. Stem 4-6 inches high, from a tuber deep in the ground, producing 2 leaves, the lower radical. Leaflets much incised. Flowers few.—Alluvial soil.

ORDER XL. ARALIA'CEÆ. (GINSENG FAMILY.)

Herbs (with us) differing from the last Order chiefly in having, as a rule, *more than 2 styles*, and the *fruit a drupe*. The umbels, also, are either single, or a cylindrical or panicled. Flowers often polygamous. The only Canadian genus is

ARALIA. GINSENG. WILD SARSAPARILLA.

* Umbels *corymbed* or panicled. *Petals, stamens, and styles each 5. Fruit black or* dark purple.

1. **A. racemosa.** (SPIKENARD.) Umbels in a large compound *panicle.* Stem 2-3 feet high, widely branching. Leaves very large and decompound : leaflets ovate-cordate, doubly serrate. Roots aromatic.—Rich woods.

2. **A. hispida.** (BRISTLY SARSAPARILLA. WILD ELDER.) Stem 1-2 feet high, *bristly*, leafy, somewhat shrubby at the base. *Umbels several, small.* Leaves twice pinnate. Leaflets sharply serrate. Fruit black.—Rocky or sandy woods.

3. **A.** nudicaulis. (WILD SARSAPARILLA.) True stem very short, sending up a naked scape bearing 3 or 4 large-stalked umbels at the summit, and one large globose leaf, ternately divided and again with 5 leaflets on each division. Root horizontal, aromatic.—Rich woods.

* * Umbel single, on a long peduncle. *Styles 2 or 3.*

4. **A.** quinquefolia. (GINSENG.) Leaves in a whorl of 3 at the summit of the stem, the latter a foot high. *Leaflets mostly 5, long-stalked.*—Rich woods.

5. **A.** trifolia. Stem 4-6 inches high. Leaves in a whorl of 3 at the summit, but the leaflets usually only 3, and sessile.—Rich woods.

ORDER XLI. CORNA'CEÆ. (DOGWOOD FAMILY.)

Shrubs or trees (rarely herbs) with simple leaves. Calyx-tube adherent to the 1–2-celled ovary, the limb of the calyx inconspicuous. Petals 4. Stamens 4, all epigynous. Style 1; stigma flat or capitate. Fruit a 1–2-seeded drupe. Flowers in cymes or in close heads surrounded by a showy involucre resembling a corolla. The only Canadian genus is

CORNUS. CORNEL. DOGWOOD.

* *Flowers in a close head, surrounded by a showy involucre of 4 white bracts. Fruit red.*

1. **C.** Canadensis. (BUNCH-BERRY.) Stem simple, 5 or 6 inches high. Upper leaves crowded and apparently whorled, ovate, the lower scale-like. Leaves of the involucre ovate. —Rich woods.

2. **C. florida.** (FLOWERING DOGWOOD.) A small tree, with opposite ovate pointed leaves. Leaves of the involucre notched at the apex.—Rocky woods. South-westward.

* * *Flowers (white) in flat cymes. No involucre. Fruit blue or white.*

3. **C.** circinata. (ROUND-LEAVED DOGWOOD.) A shrub, 4–6 feet high, with greenish warty-dotted branches. Leaves opposite, broadly oval, white-woolly beneath. Fruit light blue.— Rich woods.

4. **C.** sericea. (SILKY CORNEL.) A large shrub, with purplish branches. Leaves opposite, narrowly ovate or oblong, silky beneath. Branchlets often rusty. Fruit light blue. Distinguished from No. 3 by the colour of the branches and the much smaller leaves.—Low wet grounds.

5. **C.** stolonifera. (OSIER-LIKE DOGWOOD.) A shrub spreading freely by the production of suckers or stolons, 3–6 feet high. Branches bright red-purple, smooth. Leaves opposite, ovate, roughish whitish beneath. Fruit white or whitish.— Low wet grounds.

6. **C.** paniculata. (PANICLED CORNEL.) A shrub 4–8 feet high, with erect, gray and smooth branches. Flowers white,

very numerous. Leaves opposite, ovate-lanceolate, taper-point-
el. Cymes convex. Fruit white.—Thickets and river-banks.

7. **C. alternifolia.** (Alternate-leaved Cornel.) A large
shrub or small tree, with *alternate greenish branches streaked with
white. Leaves mostly alternate*, oval, acute at each end, crowded
at the ends of the branches. Flowers yellowish, in loose cymes.
Fruit deep blue, on reddish stalks.—Thickets.

Division II. GAMOPETALOUS EXOGENS.

Embracing plants with both calyx and corolla, the latter with the petals united (in however slight a degree).

Order XLII. **CAPRIFOLIA'CEÆ.** (Honeysuckle F.)

Shrubs, rarely herbs, with the calyx-tube adherent to the
ovary, the corolla borne on the ovary, and the stamens on
the tube of the corolla. Leaves opposite and without stipules, but some species of Vibur'num have appendages
resembling stipules. Fruit a berry, drupe, or pod.

Synopsis of the Genera.

● *Corolla tubular, sometimes 2-lipped. Style slender.*

1. Linnæa. A trailing or creeping herb, with evergreen oval crenate leaves
   and slender scaly-like peduncles which fork at the top into 2 pedicels,
   each of which bears a pair of nodding narrowly bell-shaped purple-ish
   flowers. Stamens 4. 3 shorter than the others.

2. Symphoricarpus. Upright branching shrubs, with oval entire short-
   petioled leaves. Flowers in interrupted spikes at the ends of the
   branches, crowded in 1. Corolla bell-shaped, 4-5-lobed, with as many
   stamens. Berry large and white, in clusters, her-by 2-celled.

3. Lonicera. Upright or twining shrubs, with entire leaves. Corolla
   tubular or funnel-form, irregular, *often with a* protuberance on one side
   at the base. Berry several-seeded.

4. Diervilla. Low upright shrubs with ovate pointed serrate leaves.
   Calyx-tube tapering towards the top, the teeth slender. Flowers light
   yellow, peduncles mostly 3-flowered. Corolla funnel-form, nearly regular. Pod slender, pointed.

5. Triosteum. Coarse herbs. Lobes of the calyx leaf-like. Flowers
   greenish-purple, sessile in the axils of the leaves. Corolla bulging at
   the base. Fruit a 3-seeded orange-coloured drupe.

* * *Corolla rotate or urn-shaped, regular, 5-lobed. Flowers white, in broad cymes.*

6. **Sambu'cus.** Upright shrubs with pinnate leaves, the leaflets serrate. Stigmas 3. Fruit purple or red, a juicy berry-like drupe, with 3 seed-like stones.

7. **Vibur'num.** Upright shrubs with simple leaves, and white flowers in compound cymes. Fruit a 1-seeded drupe.

<h3 style="text-align:center">1. LINNÆA. Twinflower.</h3>

L. borea lis.—Cool mossy woods and swamps.

<h3 style="text-align:center">2. SYMPHORICAR'PUS. Snowberry.</h3>

S. raceme'sus. (Snowberry.) Corolla bearded inside. Flowers in a rather loose spike.—Dry rocky hill-sides.

<h3 style="text-align:center">3. LONICE'RA. Honeysuckle. Woodbine.</h3>

1. **L. parviflo'ra.** (Small Honeysuckle.) *Twining shrub*, 2-4 feet high, with smooth leaves which are glaucous beneath, the upper ones connate-perfoliate ; corolla yellowish-purple.— Ro ky banks.

2. **L. hirsu'ta.** (Hairy Honeysuckle.) Stem *twining high*. Leaves *not glaucous, very large*, downy-hairy, the upper ones connate-perfoliate. Flowers in close whorls ; corolla greenish-yellow, clammy-pubescent.—Damp thickets.

3. **L. cilia'ta.** (Fly-Honeysuckle.) A branching upright shrub, with *thin oblong-ovate ciliate leaves*. Peduncles axillary, filiform, shorter than the leaves, each 2-*flowered at the top*. Corolla greenish-yellow, almost spurred *at the base*. The two berries separate.—Damp woods.

4. **L. oblongifolia.** (Swamp Fly-Honeysuckle.) A shrub with upright branches, and oblong *leaves*. *Peduncles long and slender, 2-flowered*. Corolla deeply 2-lipped. Berries united at the base.—Swamps and low grounds.

<h3 style="text-align:center">4. DIERVIL'LA. Bush Honeysuckle.</h3>

D. trif'ida.—Rocky woods and clearings.

<h3 style="text-align:center">5. TRIOS'TEUM. Feverwort.</h3>

T. perfolia'tum. A coarse herb, 2-4 feet high, soft-hairy. Leaves oval, narrowed at the base. Fruit orange-coloured.—Old clearings and thickets.

6. SAMBUCUS. ELDER.

1. **S. Canadensis.** (COMMON ELDER.) Shrub 5–10 feet high, in clumps. Leaflets 7–10, oblong. Cymes flat. Fruit black-purple.—Open grounds, and along streams.

2. **S. pubens** (RED-BERRIED ELDER) may be distinguished from No. 1, by i s warty bark, brown pith, 5–7 leaflets, convex or pyramidal cymes, and red berries.—Rocky woods.

7. VIBURNUM. ARROW-WOOD. LAURESTINUS.

1 **V. Lenta'go.** (SWEET VIBURNUM. SHEEP-BERRY.) A small tree, with ovate finely-serrate pointed leaves, with long and margi ed petioles. Cyme sessile. Fruit black.—Along streams.

2. **V. nudum.** (WITHE-ROD.) A smooth shrub with tall straight stems. Leaves thickish, entire or wavy-toothed, dotted beneath. *Cymes with short peduncles.* Fruit black.—Cold swamps.

3. **V. pubes'cens.** (DOWNY ARROW-WOOD.) A straggling shrub, not more than 4 feet high, with small ovate coarsely serrate leaves, *the lower surface soft downy.* Cymes small. Fruit oblong, dark-purple.—Rocky places.

4. **V. acerifo'lium.** (MAPLE-LEAVED A. DOCKMACKIE.) A shrub 3–6 feet high, with greenish bark. Leaves 3-lobed, 3-ribbed, soft-downy beneath. Stipular appendages bristle-shaped. Cymes small, on long peduncles. Fruit red, becoming black.—Thickets and river banks.

5. **V. Op'ulu?.** (CRANBERRY-TREE.) An upright shrub, 5–10 feet high, with strongly 3-lobed leaves, broader than long, the lobes spreading a d pointed. Cymes peduncled. Marginal flowers of the cyme very large and neutral. Stipular appendages conspicuous. Fruit red, pleasantly acid.—Low grounds.

6. **V. lantanoi'des.** (HOBBLE-BUSH.) A straggling shrub with reclining branches. Leaves large, round-ovate, heart-shaped at the base, serrate, many-veined, the veins underneath and the stalks and branchlets very rusty-scurfy. Stipular appendages conspicuous. Cymes sessile, very broad and flat, with very conspicuous neutral flowers on the margin. —Moist woods.

ORDER XLIII. RUBIA'CEÆ. (MADDER FAMILY.)

Herbs or shrubs, chiefly distinguished from the preceding Order by the presence of stipules between the opposite entire leaves, or by the leaves being in whorls without stipules. Calyx superior. Stamens alternate with the (mostly 5) lobes of the corolla, and inserted on its tube. Ovary 2-4-celled.

Synopsis of the Genera.

1. **Ga'lium.** *Leaves in whorls.* Slender weak herbs *with square stems. Calyx-teeth none or minute.* Corolla 4-parted, wheel-shaped. *Styles* 2. *Fruit twin, separating into two-seeded carpels.*

2. **Cepha'anthus.** *Leaves opposite.* Shrubs with the flowers in a globular *peduncled head.* Lobes of calyx and corolla each 4. Style very slender, much protruded. *Stigma capitate.*

3. **Mitchel'la.** *Leaves opposite.* Slender trailing *evergreen herbs*, with flowers in pairs, *the ovaries united.* Lobes of calyx and corolla each 4, the **corolla bearded** inside. Style 1. Stigmas 4. Fruit a red 2-eyed berry.

4. **Hous-to'nia** *Leaves* **opposite.** Low *and slender* **erect** *herbs*, with the flowers in small terminal clusters. Lobes of calyx and corolla each 4. Style 1. Stigmas 2.

1. GALIUM. BEDSTRAW. CLEAVERS.

1. **G.** aparine. (CLEAVERS. GOOSE-GRASS.) *Leaves about 8 in a whorl*, lanceolate, rough-margined. *Peduncles* 1-2-flowered, *axillary. Fruit covered with hooked prickles.* —Low grounds.

2. **G.** trilorum. (SWEET-SCENTED BEDSTRAW.) *Leaves chiefly 6 in a whorl*, elliptical-lanceolate, but thin-pointed. *Peduncles 3-flowered, terminating* the *branches.* *Fruit covered with hooked prickles.* Woods.

3. **G.** lanceolatum. (WILD LIQUORICE.) *Leaves all in whorls of 4 each*, lanceolate, *tapering to the apex*, more or less 3-nerved. *Flowers few or several, remote.* *Fruit covered with hooked prickles.*

4. G. circæzans *is similar to* No. **3, but the leaves are** *obtuse instead of tapering.* Woods.

5. G. asprellum. (ROUGH BEDSTRAW.) Leaves in whorls of **6, or** 4 or 5 on the branchlets, elliptical-lanceolate, very rough on the edges and midrib. *Stem* weak, 3-5 feet high, leaning upon

and clinging to bushes by its rough ...... *Flowers nu- paniched clusters.* Fruit not rough.—Thickets.

6. **G. trif'idum.** (SMALL BEDSTRAW.) Leaves in w... 4-6. Stem 6-18 inches high, roughened on the edges, as ar... leaves usually. *Flowers few, and panicled. Parts of the flowers generally in threes.* Fruit smooth.—Low grounds and swamps.

7. **G. borea'le.** (NORTHERN BEDSTRAW.) *Leaves in whorls of 4,* linear-lanceolate, 3-nerved. Flowers very numerous, crowded in a narrow and compact terminal panicle. Stem erect and rigid, 1-3 feet high.—Rocky thickets and river-banks.

3. CEPHALAN'THUS. BUTTON BUSH.

C. occidenta'lis. A smooth shrub in swamps, with ovate petioled pointed leaves, which are opposite or in whorls of 3. Easily recognized by the globular head of white flowers.

3. MITCHEL'LA. PARTRIDGE BERRY.

M. repens.—Common in dry woods. Leaves round-ovate, shining, sometimes with whitish lines.

4. HOUSTO'NIA. HOUSTONIA.

H. purpu'rea. Stems tufted, 3-6 inches high. Leaves vary-ing from roundish-ovate to lanceolate, 3-5-ribbed, sessile.— Woodlands.

ORDER XLIV. VALERIANA'CEÆ. (VALERIAN F.)

Herbs with opposite exstipulate leaves, and small cymose flowers. Calyx-tube adherent to the ovary, the latter 3-celled, but only one of these fertile. Stamens 1-3, fewer than the lobes of the corolla. Style slender. Stigmas 1-3. The only common genus is

ORDER XLV. **DIPSA'CEÆ.** (TEASEL FAMILY.)

Herbs with the flowers in heads, surrounded by a many-leaved involucre, as in the next Family, *but the stamens are distinct.* Leaves opposite. Represented in Canada by the genus

DIP'SACUS. TEASEL.

D. sylvestris. (WILD TEASEL.) A stout coarse prickly plant, not unlike a thistle in appearance. Flowers in oblong very dense heads, bluish. Corolla 4-cleft. Stamens 4, on the corolla. Bracts among the flowers terminating in a long awn. Leaves generally connate.—Roadsides and ditches. Rather common in the Niagara district, but found also elsewhere.

ORDER XLVI. **COMPOSITÆ.** (COMPOSITE FAMILY.)

Flowers in a dense head on a common receptacle, and surrounded by an involucre. Calyx-tube adherent to the ovary, its limb either obsolete or forming a pappus of few or many bristles or chaffy scales. Corolla either tubular or with one side much prolonged (strap-shaped or ligulate). Stamens usually 5, on the tube of the corolla, their anthers united (syngenesious). Style 2-cleft. See Part I., sections 47–49, for examination of a typical flower.)

The heads of flowers present some variety of structure. All the flowers of a head may be tubular; or only the central ones or *disk-flowers,* as they are then called, may be tubular, whilst those around the margin, then known as *ray-flowers,* are ligulate or strap-shaped. Or again, *all* the flowers may be strap-shaped. It is not unusual also to find a mixture of perfect and imperfect flowers in the same head.

The bracts which are often found growing on the common receptacle among the flowers are known as the *chaff.* When these bracts are entirely absent the receptacle is said to be naked. The leaves of the involucre are called its scales.

Artificial Synopsis of the Genera.

SUBORDER I. TUBULIFLORÆ.

Heads either altogether without strap-shaped corollas, or the latter, if present, forming only the outer circle (the *ray*). Ray flowers, when present, *always without stamens*, and often without a pistil also.

A. Ray-flowers entirely absent.

＊ Scales of the involucre in many rows, *bristly-pointed, or fringed.*

＋ *Florets all perfect.*

1. **Cir'sium.** *Leaves and scales of the involucre prickly.* Pappus of long plumose bristles. Receptacle with long soft bristles among the florets. Flowers reddish-purple.

2. **Onopor'don** *Leaves and scales of the involucre prickly.* Heads much as in Cirsium, but the *receptacle naked*, and deeply honeycombed. Pappus of long bristles, *not plumose.* Stem winged by the decurrent bases of the leaves. Flowers purple.

3. **Lap'pa.** *Leaves not prickly, but the scales of the globular involucre tipped with hooked bristles.* Pappus of many short rough bristles. Receptacle bristly. Flowers purple.

＋ ＋ *Marginal florets sterile, and their corollas much larger than the others, forming a kind of false ray.*

4. **Cen'taure'a.** Leaves not prickly. *Scales of the involucre fringed.* Pappus very short. Receptacle bristly. Flowers blue.

＋ ＋ ＋ *Sterile and fertile florets in separate heads, i.e. monœcious. Fruit a completely closed involucre outside bristly containing only one or two florets, these heads sessile in the axils of the bracts or upper leaves. Sterile heads with more numerous florets in flattish involucres, and forming racemes or spikes. Pappus none.*

5. **Xan'thium.** Fertile florets only 2 together in burs with hooked prickles, clustered in the axils. Sterile heads in short spikes above them, the scales of their involucres in one row only, but not united together.

6. **Ambro'sin.** Fertile florets single, in a closed involucre armed with a few spines at the top. Sterile heads in racemes or spikes above, the scales of their involucres in a single row and united into a cup.

＊ ＊ Scales of the involucre without bristles of any kind.

＋ Marginal **florets without stamens.**

＋ *Pappus none or minute. Receptacle naked. Very strong-scented herbs.*

7. **Tanace'tum.** Flowers yellow, in numerous corymbed heads. Scales of the involucre dry, imbricated. Pappus 5-lobed. Leaves dissected.

8. **Artemis'ia.** Flow rs yellowi h or dull purpli-h, in numerous small heads which are panicle l or racemed. Scales of the involucre with dry and scariou, margins, imbricated. Hoary herbs.

++ ++ *Pappus of all the florets bristly. Receptacle naked.*

9. **Erichthi'tes.** Flowers whitish. *Scales of the head are in a single row, linear, with a few bractlets at the base.* Corolla of the marginal florets very slender. Pappus copious, of fine soft white hairs. Heads corymbed. Erect and coarse herbs.

10. **Gnaphalium.** Flowers whitish or yellowish. Scales of the involucre yellowish-white, in many rows, dry and scarious, woolly at the base. Outer corollas slender. Pappus a single row of fine rough bristles. Flocculent woolly herbs.

11. **Antenna'ria.** Very much like Gnaphalium in appearance, being white-woolly, but the heads are nearly dioecious, and the bristles of the pappus thicker in the sterile florets.

++ ++ All the florets in the head perfect.

11. **Antenna'ria,** with dioecious heads, may be looked for here. See previous paragraph.

Bidens. One or two species have no rays. See No. 25.

Senc'cio. One species is without rays. See No. 14.

12. **Lia'tris.** Flowers handsome, rose-purple. Receptacle naked. Pappus of long and slender bristles, plumose or rou_h. Achenes slender, 10-ribbed. Lobes of the corolla slender. Stem wand-like, hairy. Leaves narrow or *grass-like*.

13. **Eupato'rium.** Flowers white or purple. Receptacle naked. Pappus of slender hair-like bristles, smooth or nearly so. Achenes 5-angled. Heads in corymbs. Leaves whorled, or connate, or opposite.

B. Rays or strap-shaped corollas round the margin of the head.

* *Pappus of hair-like bristles. Receptacle naked.*

14. **Senc'cio.** Rays yellow, or in one species none. Scales of the involucre in a single row, with a few bractlets at the base. Pappus very fine and soft. Heads corymbed. Leaves alternate.

15. **Inu'la.** Rays yellow, numerous, very narrow, in a single row. Outer scales of the involucre leaf-like. Anthers with two tails at the base. Stout plants, with large alternate leaves which are woolly beneath.

16. **Solida'go.** Rays yellow, few, as are also the disk florets. Involucre oblong, scales of unequal lengths, appressed. Achenes many-ribbed. Heads small, in compound fascicles, or corymbs. Stems usually wand-like. Leaves alternate.

17. **Aster.** Rays white, purple, or blue, *never yellow*, but the *disk* generally yellow. Pappus a *single row* of numerous fine roughish bristles. Achenes flattish. Heads corymbed or racemose. Flowering in late summer.

18. **Erig'eron.** Rays and disk as in Aster, *but the rays very narrow, and usually in more than one row.* Scales of the involucre in one or two rows, nearly of equal length. Pappus of long bristles *with shorter ones intermixed.* Heads corymbed or solitary. Leaves generally sessile.

19. **Diplopap'pus.** Rays white, long. Disk-florets yellow. Scales of the involucre 1-nerved. *Pappus double,* the outer row of short stiff bristles. Heads small, corymbed.

* * *Pappus not of hair-like* bristles, *but* either altogether wanting or consisting of *a few chaffy* scales *or teeth, or only a* minute *crown.*

⊢ Receptacle naked.

20. **Hele'nium.** Rays yellow, wedge-shaded, 3-5 cleft at the summit. Scales of the involucre reflexed, awl-shaped. Pappus of 5-8 chaffy scales, 1-nerved, the nerve usually extending into a point. Leaves alternate, decurrent on the angled stem. Heads corymbed, showy.

21. **Leucan'themum.** Rays white; disk yellow. Disk-corollas with a flattened tube. *Pappus none.* Heads single.

⊢ ⊢ Receptacle chaffy.

22. **Maru'ta.** Rays white, *s on reflexed;* disk yellow. *Ray-florets* neutral. Pappus none. Receptacle conical, more or less chaffy. Herbs with strong odour.

23. **Rudbeck'ia.** Rays yellow, usually long, *disk dark-purple, or in one species greenish yellow.* Scales of the involucre leaf-like. Receptacle conical. Pappus none, or only a minute crown. Ray-florets neutral.

24. **Helian'thus.** Rays yellow, neutral. Receptacle flattish or convex. Chaff persistent, and *embracing the 4-sided achenes.* Pappus deciduous, of 2 thin scales. Stout coarse herbs.

25. **Bidens.** Rays yellow, few; but 2 species are without rays. Scales of the involucre in 2 rows, the outer large and leaf-like. Ray-florets neutral. *Achenes crowned with 2 or more stiff awns which are barbed backward.*

26. **Achille'a.** Rays white occasionally pinkish, few. Receptacle flattish. Pappus none. Achenes margined. Heads small, in flat corymbs. Leaves very finely dissected.

27. **Polymnia.** Rays whitish-yellow wedge-form, shorter than the involucre, few in number. Scales of the involucre in 2 rows, the outer leaf-like, the inner small, and partly clasping the achenes. Pappus none. Coarse clammy herbs with an unpleasant odour.

23. **Sil'phium.** Easily known by its stout square stem, and the upper connate leaves forming a sort of cup. Flowers yellow. Achenes broad and flat.

SUBORDER II. LIGULIFLORÆ.

Corolla strap-shaped in all the florets of the head. All the florets perfect. Herbs with milky juice, and alternate leaves.

29. Lamp'sana. Flowers yellow. 8–12 in a head. Scales of the involucre 8, in a single row. *Pappus none.* Stem slender. Heads small, in loose panicles.

30. **Cicho'rium.** Flowers bright blue, showy. Scales of the involucre in 2 rows, the outer of 5 short scales, the inner of 8–10 scales. *Pappus chaffy.* Heads sessile, 2 or 3 together.

31. Leon'todon. Flowers yellow. Involucre with bractlets at the base. *Pappus of plumose bristles*, these broader at the base. Heads borne on branching scapes. Leaves radical.

32. Hiera'cium. Flowers yellow. Scales of the involucre more or less imbricated. *Pappus a single row of tawny hair-like rough bristles.* Heads corymbose.

33. Nab'alus. Flowers yellowish- or greenish-white, often tinged with purple; heads nodding. Involucre of 5–14 scales in a single row, with a few bractlets below. *Pappus copious*, of brownish or yellowish rough bristles.

34. **Tarax'acum.** Flowers yellow, on slender naked hollow scapes. Achenes prolonged into a slender thread-like beak. Leaves all radical. (See Part I., Chapter vii.)

35. Lactu'ca. Flowers pale yellow. Florets few (about 20) in the head. Scales of the involucre in 2 or more rows, unequal length. *Achenes with long thread-like body*, and a pappus of very soft white bristles. Heads numerous, panicled. Tall smooth herbs with alternate leaves.

36. Mulge'dium. Flowers bluish. Structure of the heads and general aspect of the plant as in Lactuca, *but the body of the achene short and thick and the pappus tawny.* Heads in a dense raceme.

37. Son'chus. Flowers pale yellow. Heads many-flowered, enlarging at the base. Achenes with no beak. Pappus very soft and white. Tall glaucous herbs with spiny-toothed leaves.

1. CIR'SIUM. Common Thistle.

1. **C. lanceola'tum.** (Common Thistle.) *All the scales of the involucre prickly-pointed.* Leaves decurrent, pinnatifid, the lobes prickly-pointed, rough above, woolly with webby hairs beneath.—Fields and roadsides everywhere.

2. **C. dis'color.** The *inner* scales of the involucre not prickly. Stem grooved. Leaves prickly, green above, *white-woolly beneath.* Flowers pale purple. Whole plant with a whitish aspect.—Dry thickets.

3. **C. mu'ticum.** (SWAMP THISTLE.) *Scales of the webby involucre hardly prickly*, and not *spreading. Stem very tall* and smoothish, and sparingly leafy. *Heads single or few.*—Swamps and low woods.

4. **C. arvense.** (CANADA THISTLE.) Scales of the involucre with reflexed points. Leaves prickly, smooth both sides, or slightly woolly beneath. Roots extensively creeping. Heads small and numerous.—Fields and roadsides.

2. ONOPOR'DON. SCOTCH THISTLE.

O. acan'thium. A coarse branching herb, 2–4 feet high, with woolly stem and leaves. Bristles of the pappus united at the base into a ring.—Roadsides and old fields ; not common.

3. LAP'PA. BURDOCK.

L. officina'lis. A coarse plant with very large cordate petioled leaves, and numerous small globular heads of purple flowers. The involucre forms a bur which clings to one's clothing, or to the hair of animals.—Near dwellings, mostly in manured soil.

4. CENTAURE'A. STAR-THISTLE.

C. Cy'anus. (BLUE-BOTTLE.) An old garden-plant, found occasionally along roadsides. False rays very large. Scales of the involucre fringed. Leaves linear, entire or nearly so. Stem erect. Heads single at the ends of the branches.

5. XAN'THIUM. CLOTBUR.

1. **X. struma'rium,** var. **echina'tum.** (COMMON COCKLEBUR.) Stem rough, not prickly or spiny. Leaves broadly triangular, and somewhat heart-shaped, long-petioled. Fruit a hard 2-celled bur, nearly an inch long, clothed with stiff hooked prickles, the two beaks of the fruit long and usually incurved.—Low river-banks.

2. **X.** spinosum. (SPINY CLOTBUR.) Stem armed with conspicuous straw-coloured triple slender spines, at the bases of the lanceolate short-petioled leaves, the latter white-woolly be-

neath.— Town of Dundas, Ontario; the seeds having been brought in wool from South America.

6. AMBROSIA. RAGWEED.

1. **A. artemisiæfo'lia.** (HOGWEED.) Stem erect, 1-3 feet high, branching, hairy. *Leaves tripe-pinnatifid*, the lobes linear, paler beneath.—Waste places everywhere, but not so common northward.

2. **A. trif'ida** (GREAT RAGWEED) is found in low grounds in the south-west of Ontario. Stem stouter than No. 1, 2-4 feet high. *Leaves opposite, deeply 3-cleft*, the lobes oval-lanceolate and serrate.

7. TANACETUM. TANSY.

T. vulga're. (COMMON TANSY.) A very strong-scented herb, 2-4 feet high, smooth. Leaves twice pinnate, the lobes serrate, as are also the wings of the petiole. Heads densely corymbed. — Old gardens and road-sides near dwellings.

8. ARTEMISIA. WORMWOOD.

1. **A. Canadensis.** Stem smooth or sometimes hoary with silky down, erect, usually brownish. Lower leaves twice-pinnatifid, the lobes linear.—Shores of the Great Lakes.

2. **A. vulgaris.** (COMMON MUGWORT.) Stem tall, and branching above. *Leaves green and smooth above*, white-woolly beneath, pinnatifid, the lobes linear-lanceolate. Heads small, erect in panicles. Flowers purplish.—Old fields near dwellings.

3. **A. Absin'thium.** (COMMON WORMWOOD.) Somewhat shrubby. Whole plant silky-hoary. Stem angular, branched, the branches with drooping extremities. Leaves 2-3 pinnately-divided, the lobes lanceolate. Heads nodding.—Escaped from gardens in some places.

9. ERECHTHITES. FIREWEED.

E. hieracifo'lia. Stem tall, grooved. Leaves sessile, lanceolate, cut-toothed, upper ones clasping.—Common in places recently over-run by fire.

10. GNAPHALIUM. CUDWEED.

1. G. decur'rens. (EVERLASTING.) Stem erect, 2 feet high, *clammy-pubescent*, white-woolly on the branches. Heads

corymbed.   Leaves linear-lanceolate, partly clasping, *decurrent.*
—Fields and hillsides.

2. **G. polyceph'alum.** (COMMON EVERLASTING.) Stem erect,
1-2 feet high, white-woolly.   Heads corymbed.   Leaves lanceolate, tapering at the base, *not decurrent.*—Old pastures and
woods.

3. **G. uligino'sum.** (Low CUDWEED.) Stem spreading, 3-6
inches high, white-woolly.   Leaves linear.   Heads small *in
crowded terminal clusters* subtended by leaves.—Low grounds.

11. ANTENNARIA. EVERLASTING.

1. **A. margarita'cea.** (PEARLY EVERLASTING.) Stems in
clusters, downy.   Leaves linear-lanceolate, taper-pointed, sessile.
Scales of the involucre pearly-white.   Heads in corymbs.—Along
fences and in open woods.

2. **A. plantaginifo lia.** (PLANTAIN-LEAVED E.) *Stem scape-
like, 4-6 inches high.* Radical leaves spathulate or obovate ; stem-
leaves few, linear.   Heads small, in a crowded corymb.   Involucre white or purplish.—Old pastures and woods.

12. LIATRIS. BLAZING-STAR.   •

1. **L. cylindra'cea.**   Stem wand-like, 6-18 inches high.
Leaves linear, rigid, generally 1-nerved.   Heads few, cylindrical.
—Sandy fields and thickets.

2 **L. scario'sa.**   Stem stout, 2-5 feet high.   Leaves lanceolate.   Heads very large and handsome.—Dry soil, south-
western Ontario.

3. **L. spica'ta.** Stem stout and rigid, 2-5 feet high, very
leafy.   Leaves linear, erect, the lowest 3-5 nerved.   Heads
crowded in a long spike.—Low grounds, south-western Ontario.

13. EUPATORIUM. THOROUGHWORT.

1. **E. purpu'reum.** (JOE-PYE WEED. TRUMPET-WEED.)
Stem tall and simple. *Leaves pointed, 3-6 in a whorl. Flowers
purplish or flesh-coloured.* Heads in dense corymbs.—Low
grounds.

2. **E. perfolia'tum.** (BONESET.) Stem short, hairy. *Leaves
rugose, connate-perfoliate, tapering.* Flowers whitish. Corymbs
very large.—Low grounds.

3. E. **ageratoi'des.** (WHITE SNAKE-ROOT.) Stem very smooth, commonly branching, 2-3 feet high. *Leaves opposite,* petioled, *broad'y ovate,* pointed, coarsely *serrate.* Flowers white, in corymbs.—Low rich woods.

14. SENECIO. GROUNDSEL.

1. **S. vulga'ris.** (COMMON GROUNDSEL.) *Ray-florets wanting.* Stem low, branching. Leaves pinnatifid and toothed, clasping. Flowers yellow, terminal.—Cultivated and waste grounds.

2. S. au'reus. (GOLDEN RAGWORT. SQUAW-WEED.) *Rays 8-12.* Stem smooth, or woolly when young, 1-2 feet high. Root-leaves simple, rounded, usually cordate, crenately-toothed, long-petioled. Stem-leaves sessile, lanceolate, deeply pinnatifid. Heads in a corymb nearly like an umbel.—Swamps, and often in gardens.

15. INULA. ELECAMPANE.

I. Hele'nium. (COMMON ELECAMPANE.) Stem stout, 2-5 feet high. Root-leaves very large, ovate, petioled. Stem-leaves clasping. Rays numerous, narrow.—Roadsides.

16. SOLIDAGO. GOLDEN-ROD.

* *Heads clustered in the axils of the feather-veined leaves.*

1. S. squarro'sa. Stem stout, 2-5 feet high, simple, hairy above. Scales of the involucre with reflexed herbaceous tips. Leaves large, oblong, serrate, veiny ; the lower tapering into a long-winged petiole, the upper sessile and entire. Heads in racemose clusters, the whole forming a dense, leafy, interrupted, compound spike.—Rocky woods.

2. S. bi color. Stem hoary-pubescent, usually simple. Leaves oval-lanceolate, acute at both ends ; the lower oval and tapering into a petiole, serrate. Heads in short racemes in the upper axils, the whole forming an interrupted spike or compound raceme. Rays white. The variety concolor has yellow rays.—Dry banks and thickets.

3. S. latifo'lia. Stem smooth, not angled, zigzag, 1-3 feet high. Leaves broadly ovate or oval, strongly and sharply serrate, pointed at both ends. Heads in very short axillary clusters.—Cool woods.

4. **S. cæ'sia.** Stem smooth, angled, glaucous, slender, usually branching above. Leaves smooth, lanceolate, pointed, serrate, sessile. Heads in very short clusters in the axils of the leaves.—Rich woods and hillsides.

* * *Racemes terminal, erect, loosely thyrsoid,* **not** *one-sided. Leaves feather-veined.*

5. **S. Virga-aurea**, var. **hu'milis.** Stem low, 3-6 inches high, usually smooth ; the heads, peduncles, &c., mostly glutinous. Leaves lanceolate or oblanceolate, serrate or entire, the radical ones petiolate, obtuse, and serrate at the apex.—Rocky banks ; not common.

* * * *Heads in a compound corymb terminating the simple stem, not at all racemose.*

6. **S. Ohioen'sis.** Very smooth throughout. Stem slender, reddish, leafy. Radical leaves very long (often a foot), slightly serrate towards the apex, tapering into long margined petioles ; stem-leaves oblong-lanceolate, entire, sessile.

* * * * *Heads in one-sided racemes, spreading or recurved.*

7. **S. argu'ta.** Whole plant smooth, 1-4 feet high, rigid, branching above. Lower leaves oval or elliptical-lanceolate, serrate with spreading teeth, pointed, tapering into winged and ciliate petioles ; upper ones lanceolate. Racemes very dense, naked, at length elongated and recurved. **The variety juncea** has narrower and less serrate leaves.—Woods and banks.

8. **S. Muhlenbergii.** Stem smooth, angled or furrowed. Leaves large and thin, ovate ; the upper elliptical-lanceolate. Racemes much shorter and looser than in No. 7, and the rays much larger. Moist woods and thickets.

9. **S. altis'sima.** Stem rough-hairy. Leaves ovate-lanceolate, or oblong coarsely serrate, veiny, often corrugate. Racemes panicled, spreading.—Borders of fields and copses.

10. **S. nemora'lis.** Stem minutely and closely hoary-pubescent, simple or corymbed. Leaves more or less hoary, slightly 3-nerved, obscurely serrate or entire ; the lower oblanceolate, somewhat crenate, and tapering into a petiole. Racemes numerous, dense, at length recurved, forming a large panicle.—Dry fields.

* * * * * *Racemes one-sided and recurved,* and *the leaves plainly 3-ribbed.*

11. **S. Canadensis.** Stem rough-hairy, tall and stout. Leaves lanceolate, serrate, pubescent beneath, rough above. Panicle exceedingly large —Very common along fences and in moist thickets.

12. **S.** *sero'tina.* *Stem* very *smooth, tall and stout.* Leaves lanceolate, serrate, the veins beneath pubescent. Panicle pyramidal, of many curved racemes.—Low thickets and meadows.

13. **S.** *gigante'a.* Stem smooth, stout. Leaves *lanceolate, taper-pointed,* sharply serrate, except at the base, *smooth both sides, rough-ciliate.* *Panicle large,* pubescent.—Open thickets and *meadows.*

* * * * * * *Inflorescence* a *flat-topped corymb.*

14. **S. lanceola'ta.** *Stem* pubescent above, much branched. Leaves *linear-lanceolate, the* nerves (3–5) and margins rough-*pubescent. Heads* in dense corymbed clusters, giving a decidedly *characteristic aspect* to this species.—Low river-*margins.*

17. *ASTER.* STARWORT. *ASTER.*

* *Leaves, at least the lower ones, heart-shaped and petioled.*

1. *A.* *corymbo'sus.* *Rays white or nearly* so. Heads in *corymbs. Stem slender, 1–2 feet high, zigzag.* Leaves thin, *smoothish, sharp-pointed, coarsely serrate,* all the lower ones on *slender naked petioles.—Woodlands.*

2. *A.* *macrophyl'lus.* *Rays white* or bluish. Stem *stout, 2–3 feet high. Leaves thickish, rough,* finely serrate, *the lower long-petioled. Heads in closer corymbs* than in No. 1. *Wood-banks.*

3. *A.* *azu'reus.* *Rays bright blue. Heads racemed or pani-cled. Stem roughish, corymbosely or paniculately panicled above. Leaves entire or* nearly so, *rough; the lower cordate-ovate, on long petioles; the* upper *lanceolate* or linear, *sessile. The latest flowering of our Asters.—Dry* soil.

4. *A.* *undula'tus.* *Rays bright blue.* Heads *racemed or panicled. Stem hoary with close pubescence, spreading. Leaves*

with somewhat wavy margins, ovate or ovate-lanceolate, roughish above, downy beneath ; the lowest cordate, on margined petioles ; the upper with *winged short petioles* clasping at the base, or sessile.—Dry woods.

5. **A. cordifo'lius.** Rays pale blue or nearly white. Heads *small*, profuse, panicled. Stem much branched. Leaves thin, sharply serrate, the lower on slender ciliate petioles.—Woods and along fences.

6. **A. sagittifo'lius.** Rays pale blue or purple. Heads *small*, in dense compound racemes or panicles. Stem smooth or nearly so, erect, with ascending branches. Leaves ovate-lanceolate, pointed, pubescent, the lowest on long margined petioles, the upper contracted into a winged petiole, or lanceolate or linear.— Thickets and along fences.

* * *Upper leaves all sessile or clasping by a heart-shaped base ; lower ones not heart-shaped.*

7. **A. lævis.** *Rays large*, purple or blue. Very smooth throughout. Heads in a close panicle. Leaves lanceolate or ovate-lanceolate, chiefly entire, rough on the margins, *the upper ones clasping by an auricled base.*—Dry woods.

8. **A. Novæ An'gliæ.** Rays many, narrow, violet-purple ; *heads large. Involucre of many slender equal scales,* apparently *in a single row, clammy.* Stem stout, 3-8 feet high, hairy, corymbed above. Leaves very numerous, lanceolate, entire, clasping by an auricled base, pubescent.—River-banks and borders of woods.

9. **A. puni'ceus.** *Rays long*, lilac-blue. *Scales of the involucre narrowly linear, loose.* Stem 3-6 feet high, stout, rough-hairy, *usually purple below.* Leaves oblong-lanceolate, clasping by an auricled base, sparingly serrate in the middle, rough above, smooth beneath, pointed.—Swamps ; usually clustered.

10. **A. longifo'lius.** Rays large, numerous, purplish-blue. *Scales of the involucre in several rows, linear, with awl-shaped spreading green tips. Stem smooth.* Leaves lanceolate or linear, taper-pointed, shining above. Heads solitary or few on the branchlets.—Moist thickets along streams.

* * * *None of the leaves heart-shaped ; those of the stem sessile,
tapering at the base (except in No. 11).*

11. **A.** multiflorus. Rays white. Stem pale or hoary with
minute pubescence, 1 foot high, bushy. *Leaves crowded, linear,*
with rough margins ; the upper partly clasping. Heads crowded
on the racemose branches. Scales of the involucre with spreading green tips.—Dry soil.

12. **A. Tradescan'ti.** Rays white or whitish. Scales of the
involucre narrowly linear, in 3 or 4 rows. Heads small, very
numerous, in 1-sided *close* racemes on the branches. Stem 2–4
feet high, bushy, *smooth.* Leaves linear-lanceolate, the larger
ones with a *few remote* teeth in the middle.—Moist banks.

13. A. miser. Rays pale blue or whitish. Involucre
nearly as in No. 12. Stem *more or less hairy,* much branched.
Heads small, in *loose* racemes on the spreading branches. Leaves
lanceolate, acute at each end, *sharply serrate in the middle.*—Low
grounds.

14. **A.** simplex. Rays pale blue or whitish. Scales of the
involucre linear-awl-shaped. Stem stout, smooth or nearly so,
with numerous leafy branches. Heads medium-sized, somewhat
corymbose. Leaves smooth, lanceolate, tapering at both ends,
the lower serrate.—Moist and shady banks.

15. A. tenuifolius. Rays pale blue or whitish. Scales of
the involucre linear-awl-shaped, with very slender points. Heads
medium-sized, in panicled racemes. *Leaves* few, narrowly lanceolate, tapering to a long slender point, the lower usually serrate
in the middle. Stem much branched, pubescent in lines.—Low
thickets.

16. A. ptarmicoïdes. Rays pure white. Stems clustered,
usually a foot high, each bearing a flat corymb of small heads.
Leaves linear-lanceolate, acute, rigid, entire, mostly 1-nerved,
with rough margins. Dry or gravelly hills. Our earliest Aster.

18. ERIGERON. FLEABANE.

1. E. Canadense. (HORSE-WEED. BUTTER-WEED.) Rays
white, but very inconspicuous, shorter than their tubes. Heads
very small, numerous, in panicled racemes. Stem 3–5 feet high,

erect and wand-like, bristly-hairy.   Leaves linear, mostly entire.
—Common in burnt woods and new clearings.

2. **E. bellidifolium.**   (ROBIN'S PLANTAIN.)   Rays *bluish-
purp'e*, numerous.   Heads medium-sized, few, on slender corymbose peduncles.   Stem hairy, producing offsets from the base.
Radical leaves spathulate or obovate, toothed above the middle;
stem-leaves oblong, *few*, sessile or partly clasping, entire.—
Thickets.

3. **E. Philadelphicum.**   (COMMON FLEABANE.)   Rays rose-
purp'e, very numerous and narrow.   Heads small, few, in
corymbs.   Stem hairy, with *numerous* stem-leaves.   Radical
leaves spathulate and toothed; the upper ones clasping by a
heart-shaped base, entire.—Moist grounds.

4. **E. strigo'sum.**   (DAISY FLEABANE.)   Rays white, conspicuous, numerous.   *Pappus plainly double.*   Stem and leaves
roughish with minute appressed hairs, or nearly smooth.   Lower
leaves spathulate and slender-petioled, *entire or nearly so*, the
upper lanceolate, scattered.—Dry fields and meadows.

5. **E. an'nuum.**   (LARGER DAISY FLEABANE.)   Rays white,
tinged with purple.   Pappus double.   Stem rough with spreading hairs.   Leaves coarsely toothed; the lower ovate, tapering
into a margined petiole; the upper ovate-lanceolate.   Heads
corymbed.—Fields and meadows.

19. DIPLOPAP'PUS.   DOUBLE-BRISTLED ASTER.

D. umbella'tus.   Stem smooth, leafy to the top, tall, simple.
Leaves lanceolate, long-pointed.   Heads very numerous in flat
compound corymbs.—Moist thickets.

20. HELENIUM.   SNEEZE-WEED.

H. autumna'le.   (SNEEZE-WEED.)   Stem nearly smooth.
Leaves lanceolate, toothed.   Disk globular.—Low river- and
lake-margins.

21. LEUCANTHEMUM.   OX-EYE DAISY.

L. vulga're.   (OX-EYE DAISY.   WHITE-WEED.)   Stem erect,
naked above, bearing a single large head.   Leaves chiefly cut or
cut-toothed, the lowest spathulate, the others partly clasping.—
Pas ures and o'd fields.

II. Cot'ula. (COMMON MAY-WEED.) Stem branching. **Leaves** thrice-pinnate, finely dissected.—Road-sides everywhere.

23. RUDBECKIA. CONE-FLOWER.

1. **R.** laciniata. Rays llowe, 1-2 inches long, drooping. *Disk greenish-yellow.* Stem tall, smooth, branching. Lowest leaves pinnate of 5-7 lobed leaflets; upper ones 3-5 parted, or the uppermost undivided and generally ovate. Heads terminal, long-peduncled.—Swamps.

2. **R.** hir'ta. Rays bright yellow. *Disk purplish-brown.* Stem very rough-hairy, naked above, bearing single large heads. Leaves 3-ribbed, the lowest spatulate, narrowed into a petiole, the upper ones sessile.—Meadows.

24. HELIANTHUS. SUN-FLOWER.

1. H. strumo'sus. Stem 3-4 feet high, smooth below. Leaves broadly lanceolate, rough above and whitish beneath, pointed, serrate with small appressed teeth, short-petioled. Rays about 10.—Moist copses and low grounds.

2. **H.** divarica'tus. Stem 1-4 feet high, simple or forking above. Leaves all opposite, so-like, spreading, ses-ile, *rounded* or *truncate at the base*, ovate-lanceolate, 3-nerved, long-pointed, serrate, *rough on both sides.* Heads few, on short peduncles. Rays about 12.—Open thickets and dry plains.

3. H decapet'alus. Stem 3-6 feet high, branching, smooth below, rough above. Leaves thin, green on both sides, ovate, coarsely serrate, pointed, abruptly contracted into short margined petioles. Rays usually 10.—Thickets and river-banks.

25. BIDENS. BUR-MARIGOLD.

1. **B.** frondosa. (COMMON BEGGAR-TICKS.) Rays none. Achenes flat, wedge-obovate, crowned with two slender awns with bristles pointing upward. Stem tall, branched. Leaves thin, long-petioled, pinnately 3-5 divided, the leaflets ovate-lanceolate, pointed, serrate.

2. B. connata. (SWAMP BEGGAR-TICKS.) Rays none. Achenes flat, narrowly wedge-shaped, 2-4 awned, ciliate with minute bristles pointing downwards. Stem 1-2 feet high, smooth.

Leaves lanceolate, pointed, serrate, tapering and connate at the base, the lowest often 3-parted and decurrent on the petiole.— In shallow water and low grounds.

3. B. cer'nua. (SMALLER BUR-MARIGOLD.) *Rays short, pale yellow.* Achenes flat, wedge-obovate, *4-awned, ciliate with bristles pointing downwards.* Stem nearly smooth, 5-10 inches high. Leaves all simple, lanceolate, unequally serrate, hardly connate. Heads nodding.—Wet places.

4. B. chrysanthemoides. (LARGER BUR-MARIGOLD.) *Rays an inch long, showy, golden yellow.* Achenes wedge-shaped, 2-4 awned, *bristly downwards.* Stem smooth, 6-30 inches high, erect or ascending. Leaves lanceolate, tapering at both ends, connate, regularly serrate —Swamps and ditches.

5. B. Beckii. (WATER MARIGOLD.) *Aquatic.* Stems long and slender. Immersed leaves dissected into fine hair-like divisions ; those out of water lanceolate, slightly connate, toothed. Rays showy, golden yellow, larger than the involucre. *Achenes linear, bearing 4-6 very long awns barbed towards the apex.*—Ponds and slow streams.

26. ACHILLEA. YARROW.

A. Millefolium. (MILFOIL.) Stems simple. Leaves dissected into fine divisions. Corymb flat-topped. Rays only 4 or 5, short.—Fields and along fences ; very common.

27. POLYMNIA. LEAF-CUP.

P. Canadensis. A coarse clammy-hairy herb. Lower leaves opposite, petioled, pinnatifid ; the upper alternate, angled or lobed. Heads small ; rays pale yellow.—Shaded ravines, south-westward.

28. SILPHIUM. ROSIN-PLANT.

S. perfolia'tum (CUP-PLANT) is found in south-western Ontario. Stem stout, square, 4-8 feet high. Leaves ovate, coarsely toothed, the upper ones united by their bases.

29. LAMPSANA. NIPPLE-WORT.

L. commu'nis. Very slender and branching. Leaves angled or toothed. Heads small, loosely panicled.—Borders of springs ; not common.

30. CICHO'RIUM. Succory. Cichory.

C. In'tybus. Stem-leaves oblong or lanceolate, partly clasping ; radical ones runcinate.—Roadsides and waste places.

31. LEONTODON. Fall Dandelion.

L. autumna'le. (Fall Dandelion.) Leaves lanceolate, laciniate-toothed or pinnatifid. *Scape branched.*—Roadsides and waste places ; not common.

32. HIERACIUM. Hawkweed.

1. **H. Canadense.** (Canada Hawkweed.) *Heads large.* Stem simple, leafy, corymbed, 1–3 feet high. *Peduncles downy.* Leaves ovate-oblong, with a few coarse teeth, somewhat hairy, sessile, or the uppermost slightly clasping. Achenes tapering towards the base.—Dry banks and plains.

2. **H. scabrum.** (Rough H.) *Heads small.* Stem stout, 1–3 feet high, *rough-hairy,* corymbose. *Peduncles and involucre densely clothed with dark bristles.* Achenes not tapering.—Sandy woods and thickets.

3. **H. veno'sum** (Rattlesnake-weed), with a smooth naked scape (or bearing one leaf), and a loose corymb of very slender peduncles, is found in the Niagara region.

33. NABALUS. Rattlesnake-root.

1. **N. albus.** (White Lettuce.) Heads 8-12 flowered. *Pappus deep cinnamon-coloured.* Stem 2–4 feet high, smooth and glaucous, corymbose-paniculate. Leaves triangular-halberd-shaped, or 3–5 lobed, the uppermost oblong and undivided.—Rich woods.

2. **N. altissimus.** (Tall White Lettuce.) Heads 5-6 flowered. *Pappus pale straw-coloured.* Stem taller but more slender than in No. 1, with a narrow leafy panicle at the summit.—The woods.

34. TARAXACUM. Dandelion.

T. Dens-leo'nis. (Common Dandelion.) Outer involucre reflexed. Leaves runcinate.—Fields everywhere.

35. LACTUCA. Lettuce.

L. Canadensis. (Wild Lettuce.) Heads numerous, in a long and narrow naked panicle. Stem stout, smooth, hollow,

4–9 feet high. Leaves mostly runcinate, partly clasping, pale beneath; the upper entire. Achenes longer than their beaks.— Borders of fields and thickets.

36. MULGE'DIUM. False or Blue Lettuce.

M. leucophæ'um. Stem tall and very leafy. Heads in a dense compound panicle.—Borders of damp woods and along fences.

37. SON'CHUS. Sow-Thistle.

1. **S. olera'ceus.** (Common Sow-Thistle.) Stem-leaves runcinate, slightly toothed with soft spiny teeth, clasping; *the auricles acute.*—Manured soil about dwellings.

2. **S. asper.** (Spiny-leaved S.) Leaves *hard'y lobed*, fringed with soft spines, clasping; *the auricles rounded.* Achenes *margined.*—Same localities as No. 1.

3. **S. arvensis** (Field S.), with bright yellow flowers and bristly involucres and peduncles, is found eastward.

Order XLVII. **LOBELIA'CEÆ.** (Lobelia Family.)

Herbs, with milky acrid juice, alternate leaves, and loosely racemed flowers. Corolla irregular, 5-lobed, *the tube split down one side.* Stamens 5, syngenesious, and commonly also monadelphous, *free from the corolla.* Calyx-tube adherent to the many-seeded ovary. Style 1. The only genus is

LOBE'LIA. Lobelia.

1. **L. cardina'lis.** (Cardinal-flower.) Corolla *large, deep red.* Stem simple, 2–3 feet high, smooth. Leaves oblong-lanceolate, slightly toothed. Bracts of the flowers leaf-like.—Low grounds.

2. **L. syphilit'ica.** (Great Lobelia.) Corolla *rather large, light blue.* Stem hairy, simple, 1–2 feet high. Leaves thin, acute at both ends, serrate. Calyx-lobes half as long as the corolla, the tube hemispherical. Flowers in a dense spike or raceme.—Low grounds.

3. **L. infla'ta.** (Indian Tobacco.) Flowers *small, ⅛ of an inch long, pale blue.* Stem leafy, *branching,* 8–18 inches high, pubescent. Leaves ovate or oblong, toothed. *Pods inflated. Racemes leafy.*—Dry fields.

4. L. spica'ta. Flowers small, $\frac{1}{3}$ *of an inch long, pale blue.* Stem slender, erect, *simple,* 1–3 feet high, minutely pubescent below. Leaves barely toothed, the lower spathulate or obovate, the upper reduced to linear bracts. *Raceme long and naked.*— Sandy soil.

5. L. Kal'mii.—Flowers small, $\frac{1}{3}$ of an inch long, *light blue. Stem low,* 4–18 inches high, *very slender.* Pedicels filiform, as long as the flowers, with 2 minute bractlets above the middle. Leaves mostly linear, the radical ones spathulate and the uppermost reduced to bristly bracts.—Wet rocks and banks, chiefly northward.

6. L. Dortman'na (WATER LOBELIA), with small leaves all tufted at the root, and a scape 5 or 6 inches long with a few small light-blue pedicelled flowers at the summit, occurs in the shallow borders of ponds in Muskoka.

ORDER XLVIII. CAMPANULA'CEÆ. (CAMPANULA F.)

Herbs with milky juice, *differing from the preceding Order chiefly in having a regular 5-lobed corolla (bell-shaped or wheel-shaped), separate stamens (5), and 2 or more (with us, 3) stigmas.*

Synopsis of the Genera.

1. **Campan'ula.** Calyx 5-cleft. Corolla generally bell-shaped, 5 lobed. Pod short.
2. **Specula'ria.** Calyx 5-cleft. Corolla nearly wheel-shaped, 5-lobed. Pod prismatic or oblong.

1. CAMPANULA. BELL-FLOWER.

1. C. rotundifo'lia. (HAREBELL.) Flowers blue, loosely panicled, on long slender peduncles, nodding. Stem slender, branching, several-flowered. Root-leaves round-heart-shaped ; stem-leaves linear. Calyx-lobes awl-shaped.—Shaded cliffs.

2. C. aparinoi'des. (MARSH BELLFLOWER.) Flowers *white* or nearly so, about $\frac{1}{3}$ of an inch long. *Stem very slender and weak,* few-flowered, *angled, roughened backwards.* Leaves linear-lanceolate. Calyx-lobes triangular.—Wet places in high grass. The plant has the habit of a *Galium.*

3 C. America'na. (TALL BELLFLOWER.) Flowers light blue, about an inch across, *crowded in a leafy spike.* Corolla deeply

5-lobed. *Style long and curved.* Stem 3–6 feet high, simple. Leaves ovate or ovate-lanceolate, taper-pointed, serrate.—Moist rich soil.

2. SPECULARIA. Venus's Looking-glass.

S. perfolia'ta. Flowers purplish-blue, only the later or upper ones expanding. Stem hairy, 3–20 inches high. Leaves round-ish or ovate, clasping. Flowers solitary or 2 or 3 together in the axils.—Sterile open ground, chiefly south-westward.

Order XLIX. ERICA'CEÆ. (Heath Family.)

Chiefly shrubs, *distinguished by the anthers opening, as a rule, by a pore at the top of each cell.* Stamens (as in the two preceding Orders) free from the corolla, as many or twice as many as its lobes. Leaves simple and usually alternate. Corolla in some cases polypetalous.

Synopsis of the Genera.

Suborder I. VACCINIEÆ. (Whortleberry Family.)

Calyx-tube adherent to the ovary.   Fruit a berry crowned with the calyx-teeth.

1. **Gaylussa'cia.** Stamens 10, the anthers opening by a pore at the apex. Corolla tubular, ovoid, the border 5-cleft. *Berry 10-celled,* seeded. Flowers white with a red tinge. *Leaves covered with resinous dots.* Branching shrubs.

2. **Vaccin'ium.** Stamens 8 or 10, the anthers prolonged upward into tubes with a pore at each apex. Corolla deeply 4-parted and revolute, or cylindrical with the limb 5-toothed. Berry 4-celled, or more or less completely 10-celled. Flowers white or reddish, solitary or in short racemes. Shrubs.

3. **Chiogenes.** Stamens 8, *each anther 2-pointed at the apex.* Corolla bell-shaped, deeply 4-cleft. Limb of the calyx 4-parted. Flowers very small, nodding from the axils, with 2 bractlets under the calyx. *Berry white,* 4-celled. A trailing slender evergreen.

Suborder II. ERICINEÆ. (Heath Family Proper.)

Calyx free from the ovary.   Shrubs or small trees.   Corolla gamopetalous, except in No. 10.

4. **Arcto-staph'ylos.** Corolla urn-shaped, the limb 5-toothed, revolute. Stamens 10, the anthers each with 2 reflexed awns on the back. Fruit a berry-like drupe, 5–10 seeded. A trailing thick-leaved evergreen, with nearly white flowers.

5. **Epigæ'a.** Cor lla salver-shaped, hairy inside, rose coloured. Stamens 10; filaments slender ; anthers awnless, *opening by a Pore*. Calyx of 5 pointed and scale-like nearly dis inct sepals. A trailing evergreen, bristly with rusty hai s.

6. **Gaultheria.** Corolla ovoid, or globish urn-shaped, 5-toothed, nearly white. Stamens 10. the anthers awnless. Calyx 5-cleft. enclosing the pod and *becoming fleshy and berry-li e in fruit.* Stems low and slender, leafy at the summit.

7. **Cassan'dra.** Corolla cylindrical, 5-toothed. Stamens 10, the anther-cells tapering into beaks with a pore at the apex, awnless. Calyx of 5 overlapping sepals, and 2 similar bractlets. Pod with a double pericarp, the outer of 5 valves. the inner cartilaginous and of 10 valves. A low shrub with rather scurfy leaves, and white flowers.

8. **Androm'eda.** Corolla globular urn-shaped, 5-toothed. Calyx of 5 nearly distinct tubular sepals, without bractlets. Stamens 10 ; the filaments bearded the anther-cells each with a slender awn. A low shrub, with white flowers in a terminal umbel.

9. **Kal mia.** Corolla broadly bell-shaped, *with 10 pouches receiving as many anthers.* Shrubs with showy rose-purple flowers.

10. **Ledum.** Calyx 5-toothed. very small. Corolla of 5 obovate and spreading distinct *petals.* Stamens 5–10. Leaves evergreen, *with revolute margins,* green above and with *rusty wool.*

SUBORDER III. **PYROLEÆ.** (PYROLA FAMILY.)

Calyx free from the ovary. *Corolla polypetalous.* More or less herbaceous evergreens.

11. **Pyrola.** Calyx 5-parted. Petals 5, concave. Stamens 10. Stigma 5-lobed. Leaves evergreen, *clustered at the base of an upright scaly-bracted scape which bears a simple raceme of nodding flowers.*

12. **Moneses.** Petals 5, orbicular spreading. Stamens 10. Stigma large, peltate, with 5 narrow radiating lobes. Plant having the aspect of a Pyrola, but the scape bearing a single terminal flower.

13. **Chimaphila.** Petals 5, concave orbi ular spreading. Stamens 10. Stigma almost sessile, the disk 5-crenate. Low plants with running underground stems, the evergreen shining, sharply toothed somewhat whorled leaves. Flowers corymbose, umbelled on a terminal peduncle.

SUBORDER IV. **MONOTROPEÆ.** (INDIAN-PIPE FAMILY.)

14. **Monot'ropa.** A low, simple, white, tawny or reddish plant, having scales instead of leaves, and bearing a flower at the summit of the stem.

1. GAYLUSSACIA. HUCKLEBERRY.

G. resino'sa. (BLACK HUCKLEBERRY.) Fruit *black,* without a *bloom.* Racemes short, 1-sided. in clusters. Leaves oval or oblong. Branching shrub 1-3 feet high. —Low grounds.

2. VACCIN'IUM. CRANBERRY. BLUEBERRY.

1. **V. Oxycoc'cus.** (SMALL CRANBERRY.)   A creeping or trailing very slender shrubby plant, with ovate acute evergreen leaves only ¼ of an inch long, the margins revolute. Corolla rose-coloured, 4-parted, the lobes reflexed. Anthers 8. *Stem 4-9 inches long. Berry only about ¼ of an inch across*, often speckled with white.—Bogs.

2. **V. macrocar'pon.** (LARGE or AMERICAN CRANBERRY.) Different from No. 1 in having prolonged stems (1–3 feet long) and the flowering branches lateral. The leaves also are nearly twice as large, and the *berry is fully ½ an inch broad.*—Bogs.

3. **V. Pennsylva'nicum.** (DWARF BLUEBERRY.) Stem 6–15 inches high, the branches green, angled, and warty. Corolla cylindrical bell-shaped, 5-toothed. Anthers 10. Flowers in short racemes. Leaves lanceolate or oblong, serrulate with bristly-pointed teeth, smooth and shining on both sides. Berry blue or black with a bloom.—Dry plains and woods.

4. **V. corymbo'sum** (SWAMP-BLUEBERRY) is a tall shrub (3–10 feet) growing in swamps and low grounds, with leaves varying from ovate to elliptical-lanceolate, and flowers and berries very much the same as those of No. 3; but the berries ripen later.

3. CHIOG'ENES. CREEPING SNOWBERRY.

C. hispid'ula. Leaves very small, ovate and pointed, on short petioles, the margins revolute. The lower surface of the leaves and the branches clothed with rusty bristles. Berries bright white.— Bogs and cool woods.

4. ARCTOSTAPH'YL S. BEARBERRY.

A. Uva-ursi. Flowers in terminal racemes. Leaves alternate, obovate or spathulate, entire, smooth. Berry red.—Bare hillsides.

5. EPIG.EA. GROUND LAUREL. TRAILING ARBUTUS.

E. re'pens. (MAYFLOWER.) Flowers in small axillary clusters from scaly bracts. Leaves evergreen, rounded and heart-shaped, alternate, on slender petioles. Flowers very fragrant.— Dry woods, in early spring.

6. GAULTHE'RIA. AROMATIC WINTERGREEN.

G. procum'bens. (TEABERRY. WINTERGREEN.) Flowers mostly single in the axils, nodding. Leaves obovate or oval, obscurely serrate, evergreen. Berry bright red, edible.—Cool woods, chiefly in the shade of evergreens.

7. CASSAN'DRA. LEATHER LEAF.

C. calycula'ta. Flowers in 1-sided leafy racemes. Leaves oblong, obtuse, flat.—Bogs.

8. ANDROM'EDA. ANDROMEDA.

A. polifo'lia. Stem smooth and glaucous, 6–18 inches high. Leaves oblong-linear, with strongly revolute margins, white beneath.—Bogs.

9. KAL'MIA. AMERICAN LAUREL.

K. glau'ca. (PALE LAUREL.) A straggling shrub about a foot high, with few-flowered terminal corymbs. Branchlets 2-edged. Leaves opposite, oblong, the margins revolute. Flowers ½ an inch across.—Bogs.

10. LE'DUM. LABRADOR TEA.

L. latifo'lium. Flowers white, in terminal umbel like clusters. Leaves elliptical or oblong. Stamens 5, or occasionally 6 or 7.—Bogs.

11. PY'ROLA. WINTERGREEN. SHIN-LEAF.

1. **P. rotundifo'lia.** *Leaves orbicular, thick, shining, usually shorter than the petiole.* Calyx-lobes lanceolate. Flowers white or in one variety rose-purple.—Moist woods.

2. **P. ellip'tica.** (SHIN-LEAF.) *Leaves elliptical, thin, dull, usually longer than the margined petiole.* Flowers greenish-white. —Rich woods.

3. **P. secunda.** Easily recognized by the *flowers of the dense raceme being all turned to one side.* Style long, protruding.— Rich woods.

4. **P. chloran'tha** has small roundish dull leaves, converging greenish-white petals, and *the anther cells contracted below the pore into a distinct neck or horn.*—Open woods.

12. MONE'SES. ONE-FLOWERED PYROLA.

M. uniflo'ra. Leaves thin, rounded, veiny, and serrate. Scape 2–4 inches high, bearing a single white or rose-coloured flower.—Deep woods.

13. CHIMAPHILA. PIPSISSEWA.

C. umbella'ta. (PRINCE'S PINE.) Leaves wedge-lanceolate, acute at the base. Peduncles 4–7-flowered. Corolla rose- or flesh-coloured.—Dry woods.

14. MONOTROPA. INDIAN PIPE. PINE-SAP.

M. uniflo'ra. (INDIAN PIPE. CORPSE-PLANT.) Smooth, waxy-white, turning black in drying.—Dark rich woods.

ORDER L. AQUIFOLIACEÆ. (HOLLY FAMILY.)

Shrubs or small trees, with small axillary polygamous or diœcious flowers, the parts mostly in fours or sixes. Calyx very minute, free from the ovary. Stamens alternate with the petals, attached to their base, the corolla being almost polypetalous. Anthers opening lengthwise. Stigma nearly sessile. Fruit a berry-like 4–8-seeded drupe.

1. ILEX. HOLLY.

I. verticilla'ta. (BLACK ALDER. WINTERBERRY.) A shrub with the greenish flowers in sessile clusters, or the fertile ones solitary. Parts of the flower mostly in sixes. Fruit bright red. Leaves alternate, obovate, oval, or wedge-lanceolate, pointed, veiny, serrate.—Swamps and low grounds.

2. NEMOPANTHES. MOUNTAIN HOLLY.

N. Canadensis. A branching shrub, with grey bark, and alternate oblong nearly entire smooth leaves on slender petioles. Flowers on long slender axillary peduncles, mostly solitary. Petals 4–5, oblong-linear, *distinct.* Fruit light red.—Moist woods.

ORDER LI. PLANTAGINA'CEÆ. (PLANTAIN FAMILY.)

Herbs, with the leaves all radical, and the flowers in a close spike at the summit of a naked scape. Calyx of 4 sepals, persistent. Corolla 4-lobed, thin and membrana-ceous, spreading. Stamens 4, usually with long filaments, inserted on the corolla. Pod 2-celled, the top coming off like a lid. Leaves ribbed. The principal genus is

PLANTAGO. PLANTAIN. RIB-GRASS.

1. P. major. (Common P.) Spike long and slender. Leaves 5-7 ribbed, ovate or slightly heart-shaped, with channelled petioles.—Moist ground, about dwellings.

2. **P. lanceola′ta.** (RIB-GRASS. ENGLISH PLANTAIN.) Spike thick and dense, short. Leaves 3-5 ribbed, lanceolate or lanceo-late-oblong. Scape grooved, long and slender.—Dry fields and banks.

3. **P. marit′ima**, var. **juncoides**, with very narrow and slender spike, and linear fleshy leaves, is found on the sea-coast and **Lower St. Lawrence.**

ORDER LII. **PRIMULA′CEÆ.** (PRIMROSE **FAMILY.**)

Herbs with regular perfect flowers, well marked by having a stamen before each petal or lobe of the corolla, and inserted on the tube. Ovary 1-celled, the placenta rising from the base. Style 1; stigma 1.

Synopsis of the Genera.

1. Prim′ula. Leaves all in a cluster at the root. Flowers in an umbel at the summit of a simple scape. Corolla salver-shaped. Stamens 5 included.

2. Trienta′lis. Leaves in a whorl at the summit of a slender stem. Calyx usually 7-parted, the lobes pointed. Corolla wheel-shaped, spreading, without a tube. Filaments united in a ring below. Flowers usually only one, white and star-shaped.

3. Lysimach′ia. Leafy-stemmed. Flowers yellow, solitary, or in a loose raceme. Calyx usually 5-parted. Corolla wheel-shaped, mostly 5-parted, and sometimes 5-spurred cleft.

4. Anagal′lis. Low and spreading. Leaves opposite or whorled, entire. Flowers variously coloured, solitary in the axils. Calyx 5-parted. Corolla wheel-shaped, 5-parted. Filaments bearded.

5. Samo′lus. Smoothish herb, with a branched stem. Corolla bell-shaped, 5-parted with 5 sterile filaments in the sinuses. Pod 5-valved at the summit. Flowers very small, white, racemed. Leaves alternate.

1. PRIMULA. Primrose. Cowslip.

1. P. farino'sa. (BIRD'S-EYE P.) Lower surface of the leaves covered with a white mealiness. Corolla lilac with a yellow centre.--Shore of Lake Huron, and northward.

2. P. Mistassin'ica. Leaves not mealy. Corolla flesh coloured, the lobes obcordate.—Shores of the Upper Lakes, and northward.

2. TRIENTALIS. Chickweed-Wintergreen.

T. America'na. (STAR-FLOWER.) Leaves thin and veiny, lanceolate, tapering towards both ends. Petals pointed.—Moist woods.

3. LYSIMACHIA. Loosestrife.

1. L. thyrsiflo'ra. (TUFTED LOOSESTRIFE.) Flowers in spike-like clusters from the axils of a few of the upper leaves. Petals lance-linear, *purplish-dotted*, as many minute teeth between them. Leaves scale-like below, the upper lanceolate, opposite, sessile, dark-dotted.—Wet swamps.

2. L. stricta. Flowers on slender pedicels *in a long terminal raceme.* Petals lance-oblong, streaked with dark lines. *Leaves opposite*, lanceolate, acute at each end, sessile, dark-dotted.— Low grounds.

3. L. quadrifolia. *Flowers on long slender peduncles from the axils of the upper leaves.* Petals streaked. *Leaves in whorls* of 4 or 5, ovate-lanceolate, dark-dotted.—Sandy soil.

4. L. ciliata. Flowers nodding on slender peduncles from the upper axils. *Petals not streaked or dotted.* Leaves opposite, *not dotted*, ovate-lanceolate, pointed, cordate at the base, *on long fringed petioles.*—Low grounds.

5. L. longifo'lia. Petals not streaked or dotted. *Stem-leaves sessile, narrowly linear, 2-4 inches long,* the margins sometimes revolute. Stem 4-angled.—Moist soil, chiefly northward.

4. ANAGALLIS. Pimpernel.

A. arvensis. (COMMON PIMPERNEL.) Petals obovate, *fringed with minute teeth,* mostly bluish or purplish. Flowers closing at the approach of rain. Leaves ovate, sessile.—Sandy fields and garden soil.

S. SAMOLUS.   WATER PIMPERNEL. BROOKWEED.

S. Valeran'di, var. Americanus. Stem slender, diffusely branched. The slender pedicels each with a bractlet at the middle.—Wet places, chiefly eastward.

ORDER LIII. LENTIBULACEÆ. (BLADDERWORT F.)

Small aquatic or marsh herbs, with a 2-lipped calyx and a personate corolla with a spur or sac underneath. Stamens 2. Ovary as in Primulaceæ. Chiefly represented by the genus

UTRICULA'RIA.   BLADDERWORT.

1. **U.** vulga'ris. (GREATER BLADDERWORT.) Immersed leaves crowded, finely dissected into capillary divisions, furnished with small air-bladders. Flowers yellow, several in a raceme on a naked scape. Corolla closed; the spur conical and shorter than the lower lip.—Ponds and slow waters.

2. **U. interme'dia.** Immersed leaves 4 or 5 times forked, the divisions linear-awl-shaped, minutely bristle-toothed on the margin, *not bladder-bearing*, the bladders being on leafless branches. Stem 3-6 inches long. Scape very slender, 2-6 inches long, bearing few yellow flowers. Upper lip of the corolla much longer than the palate; *the spur closely pressed to the broad lower lip*.—Shallow waters.

3. U. cornu'ta, with an awl-shaped spur turned downward and outward, and the lower lip of the corolla helmet-shaped, is found towards the sea-coast. Flowers yellow. Leaves awl-shaped.

ORDER LIV. OROBANCHACEÆ. (BROOM-RAPE F.)

Parasitic herbs, destitute of green foliage. Corolla more or less 2-lipped. Stamens didynamous. Ovary 1-celled with 2 or 4 parietal placentæ, many-seeded.

1. EPIPHEGUS.   BEECH-DROPS.

E. Virginia'na. A yellowish-brown branching plant, parasitic on the roots of beech trees. Flowers racemose or spiked; the upper sterile, with long corolla; the lower fertile, with short corolla.

2. CONOPHOLIS. Squaw-root.

C. America'na. A chestnut-coloured or yellow plant found in clusters in oak woods in early summer, 3-6 inches high and rather less than an inch in thickness. The stem covered with fleshy scales so as to resemble a cone. Flowers under the upper scales ; stamens projecting.

Order LV. SCROPHULARIA'CEÆ. (Figwort F.)

Herbs, distinguished by a 2-lipped or more or less irregular corolla, stamens usually 4 and didynamous, or only 2, (or in Verbascum 5,) and a 2-celled and usually many-seeded ovary. Style 1 ; stigma entire or 2-lobed.

Synopsis of the Genera.

• *Corolla wheel-shaped, and only slightly irregular.*

1. **Verbas'cum.** *Stamens (with anthers) 5.* Flowers in a long terminal spike. Corolla 5-parted, nearly regular. Filaments (or some of them) woolly.

2. **Veron'ica.** *Stamens only 2 ;* filaments long and slender. Corolla mostly 4-parted, nearly or quite regular. Pod flattish. Flowers solitary in the axils, or forming a terminal raceme or spike.

• • *Corolla 2-lipped, or tubular and irregular.*

← *Upper lip of the corolla embraces the lower in the bud, except occasionally in Mimulus.*

3. **Linaria.** Corolla personate (Fig. 142, Part I.) with *a long spur beneath.* Stamens 4. Flowers yellow, in a crowded raceme.

4. **Scrophula'ria.** Corolla tubular, somewhat inflated, 5-lobed ; the 4 upper lobes erect, the lower one spreading. Stamens with anthers 4, the rudiment of a fifth in the form of a scale on the upper lip of the corolla. Flowers small and dingy, forming a narrow terminal panicle. Stem 4-sided.

5. **Chelo'ne.** Corolla inflate-tubular (Fig. 142, Part I.) Stamens 4, with woolly filaments and anthers, and a fifth filament with no an anther. Flowers white, in a close terminal spike.

6. **Pentste'mon.** Corolla 2-lipped, gradually enlarged upwards. Stamens 4, with a fifth sterile filament, *the latter yellow-bearded.* Flowers white or purplish in a loose panicle.

7. **Mimulus.** Calyx 5-angled and 5-toothed. Upper lip of the corolla erect or a-lobed spreading, the lower spreading 3-lobed. Stamens 4 under, no *rudiment of a fifth.* Stigma 2-lipped. Flowers blue or yellow, solitary on axillary peduncles.

8. **Grati'ola.** Corolla tubular and 2-lipped. *Stamens with anthers only 2*, included. Flowers with a yellowish tube, on axillary peduncles, solitary. Style dilated at the apex.

    ← ← *Lower lip of the corolla embracing the upper in the bud.*

9. **Gerar'dia.** Corolla funnel-form, swelling above, *the 5 spreading lobes more or less unequal*. Stamens 4, *strongly didynamous*, hairy. Style long, enlarged at the apex. Flowers purple or yellow, solitary on axillary peduncles, or sometimes forming a raceme.

10. **Castille'ia.** Corolla tubular and 2-lipped, *its tube included in the tubular and flattened calyx*; the upper lip long and narrow and flattened laterally, the lower short and 3 lobed. Stamens 4, didynamous. *Floral leaves scarlet* in our species. Corolla pale yellow.

11. **Euphra'sia.** *Calyx 4-cleft.* Upper lip of the corolla erect, the lower spreading. Stamens 4, under the upper lip.—Very small herbs, with whitish or bluish spiked flowers. (Chiefly on the sea-coast, and north of Lake Superior.)

12. **Rhinan'thus.** *Calyx flat, nearly inflated in fruit*, 4-toothed. Upper lip of the corolla arched, flat, with a minute tooth on each side below the apex. Stamens 4. Flowers yellow, solitary in the axils, nearly sessile, the whole forming a crowded 1-sided spike. (Chiefly on the sea-coast, and north of Lake Superior.)

13. **Pedicula'ris.** Calyx split in front, not inflated in fruit. Corolla 2-lipped, the upper lip arched or hooded, incurved, flat, 2-toothed under the apex. Stamens 4. *Pod flat, somewhat sword-shaped.*

14. **Melampy'rum.** *Calyx 4-cleft, the lobes sharp-pointed.* Corolla greenish-yellow; upper lip arched, compressed, the lower 3-lobed at the apex. Stamens 4; anthers hairy. Pod 1-4 seeded, flat, oblique. Upper leaves larger than the lower ones and fringed with bristly teeth at the base.

1. VERBASCUM. MULLEIN.

1. **V. Thap'sus.** (COMMON MULLEIN.) A tall and very woolly herb, with the simple **stem winged by the decurrent bases** of the leaves. Flowers **yellow, forming a dense** spike.—Fields and roadsides everywhere.

2. **V. Blatta'ria.** (MOTH M.) Stem slender, nearly smooth. Lower leaves petioled, doubly serrate; the upper partly clasping. Flowers whitish with a purple tinge, in a loose raceme. *Filaments all violet-bearded.*—Road-sides; not common northward.

2. VERONICA. SPEEDWELL.

1. **V. America'na.** (AMERICAN BROOKLIME.) Flowers pale blue, in *opposite* axillary racemes. *Leaves nearly petioled,*

thickish, serrate. *Pod swollen.*—A common plant in brooks and ditches.

2. V. Anagallis (WATER SPEEDWELL) is much like No. 1, but the *leaves are sessile* with a heart-shaped base.

3. V. scutellata (MARSH S.) Flowers pale blue, in racemes chiefly from alternate axils. *Leaves* sessile, linear, opposite, hardly toothed. Racemes 1 or 2, slender and strong. *Flowers* few. Pods very flat, notched at both ends.—Bogs.

4. V. officinalis. (COMMON S.) Flowers light blue. Stem prostrate, rooting at the base, pubescent. Leaves short-petioled, obovate-elliptical, serrate. *Racemes* dense, chiefly from alternate axils. Pod obovate-triangular, notched.—Hillsides and open woods.

5. V. serpyllifolia. (THYME-LEAVED S.) Flowers whitish or pale blue, in a loose *terminal* raceme. Stem nearly smooth, branched at the creeping base. Leaves obscurely crenate, the *lowest petioled.* Pod flat, notched.—Hillsides and fields. Plant only 2 or 3 inches high.

6. V. peregrina. (NECKWEED.) Flowers whitish, *solitary in the axils of the upper leaves.* Whitish corolla shorter than the calyx. Stem 4-9 inches high, nearly smooth. *Pod orbicular,* slightly notched.—Waste places and cultivated grounds.

7. V. arvensis. (CORN SPEEDWELL.) Flowers (blue) as in No. 6, but the *stem is hairy,* and the *pod inversely heart-shaped.* —Cultivated soil.

3. LINARIA. TOAD-FLAX.

L. vulgaris. (TOAD-FLAX. BUTTER AND EGGS.) Leaves crowded, linear, pale green. Corolla pale yellow, with a deeper yellow or orange-colored palate.—Roadsides.

4. SCROPHULARIA. FIGWORT.

S. nodosa. Stem smooth, 3-4 feet high. Leaves ovate or oblong, the upper lanceolate, serrate.—Damp thickets.

5. CHELONE. TURTLE-HEAD.

C. glabra. Stem smooth, erect and branching. Leaves short-petioled, lance-oblong, serrate, opposite. Bracts of the flowers concave.—Wet places.

6. PENTSTEMON. BEARD-TONGUE.

P. **pubes'cens.** Stem 1-3 feet high, pubescent ; the panicle more or less clammy. Throat of the corolla almost closed. Stem-leaves lanceolate, clasping. — Dry soil.

7. MIMULUS. MONKEY-FLOWER.

1. M. rin'gens. Stem square, 1-2 feet high. *Corolla blue*, an inch long. Leaves oblong or lanceolate, clasping. — Wet places.

2. **M.** James'ii. Stem creeping at the base. *Corolla yellow*, small. Leaves roundish or kidney-shaped, nearly sessile. Calyx inflated in fruit. — In cool springs.

8. GRATIOLA. HEDGE-HYSSOP.

G. Virginia'na. Stem 4-6 inches high, clammy with minute pubescence above. Leaves lanceolate. Peduncles slender. — Moist places.

9. GERARDIA. GERARDIA.

1. **G. purpu'rea.** (PURPLE GERARDIA.) Corolla rose-purple. Leaves linear, acute, rough-margined. *Flowers an inch long, on short peduncles.* — Low grounds.

2. **G.** tenuifo'lia. (SLENDER G.) Corolla rose-purple. Leaves linear, acute. *Flowers about ½ an inch long, on long thread-like peduncles* — Dry woods.

3. G. fla'va. (DOWNY G.) Corolla yellow, woolly inside. Stem 3-4 feet high, finely pubescent. Leaves oblong or lance-shaped, the upper entire, the lower usually more or less pinnatifid, downy-pubescent. — Woods.

4. **G.** quercifo'lia. (SMOOTH G.) Corolla yellow, woolly inside. Stem 3-6 feet high, smooth and glaucous. Lower leaves twice-pinnatifid, the upper pinnatifid or entire, smoothish. — Woods.

5. **G.** pedicula'ria. (FERN-LEAVED G.) Nearly smooth. Flowers nearly as in Nos. 3 and 4. Stem 2-3 feet high, very leafy, much branched. Leaves pinnatifid, *the lobes cut and toothed.* — Thickets.

10. CASTILLEIA. PAINTED-CUP.

C. coccin'ea. (SCARLET PAINTED-CUP.) Calyx 2-cleft, yellowish. Stem pubescent or hairy, 1-2 feet high. The stem-

leaves nearest the flowers 3-cleft, the lobes toothed, *bri* .
—Sandy soil.

11. EUPHRA'SIA. EYEBRIGHT.

E. officina'lis is rather common on the Lower St. Law
and the sea-coast. Lowest leaves crenate, those next the flowers
bristly-toothed.

12. RHINAN'THUS. YELLOW-RATTLE.

R. Crista-galli. (COMMON YELLOW-RATTLE.) Localities
much the same as those of Euphrasia. Seeds broadly winged,
rattling in the inflated calyx when ripe.

13. PEDICULARIS. LOUSEWORT.

P. Canadensis. (COMMON LOUSEWORT. WOOD BETONY.)
Stems clustered, simple, hairy. Lowest leaves pinnately parted.
Flowers in a short spike.—Copses and banks.

14. MELAMPYRUM. COW-WHEAT.

M. America'num. Leaves lanceolate, short-petioled; the
lower ones entire.—Open woods.

ORDER LVI. VERBENA'CEÆ. (VERVAIN FAMILY.)

Herbs (with us), with opposite leaves, didynamous stamens,
and corolla either irregularly 5-lobed or 2-lipped. Ovary in
Verbena 4-celled (when ripe splitting into 4 nutlets) and in
Phryma 1-celled, *but in no case 4-lobed*, thus distinguishing
the plants of this Order from those of the next.

Synopsis of the Genera.

1. **Verbena.** Flowers in spikes. Calyx tubular, 5-ribbed. Corolla tubular,
   salver-form, the border rather irregularly 5-cleft. Fruit splitting into
   4 nutlets.

2. **Phryma.** Flowers in loose slender spikes, reflexed in fruit. Calyx
   cylindrical. 2-lipped, the upper lip of 3 slender teeth. Corolla 2-lipped.
   Ovary 1-celled and 1-seeded.

1. VERBENA. VERVAIN.

1. **V.** hasta'ta. (BLUE VERVAIN.) Stem 3–5 feet high. Leaves
oblong-lanceolate, taper-pointed, serrate. Spikes of *purple* flowers dense, erect, corymbed or panicled.—Low meadows and fields.

2. **V. urticifo'lia.** (NETTLE-LEAVED V.) Stem tall. Leaves oblong-ovate, acute, coarsely serrate. Spikes of small *white* flowers very slender, loosely panicled.—Fields and roadsides.

2. PHRYMA. LOPSEED.

P. Leptostachya. Corolla purplish or pale rose-coloured. Stem slender and branching, 1–2 feet high. Leaves ovate, coarsely toothed.—Woods and thickets.

ORDER LVII. LABIATÆ. (MINT FAMILY.)

Herbs with square stems, opposite leaves (mostly aromatic), didynamous (or in one or two genera *diandrous*) stamens, a 2-lipped or irregularly 4 or 5-lobed corolla, and a deeply 4-lobed ovary, forming in fruit 4 nutlets or achenes. (See Part I., Section 50, for description of a typical plant.)

Synopsis of the Genera.

* Stamens 4, curved upwards, parallel, or arched from a deep notch on the upper side of the 2-lobed corolla.

1. **Teucrium.** Calyx 5-toothed. The 4 upper lobes of the corolla nearly equal, with a deep notch *between the upper 2*; the lower lobe much larger. Flowers dull purple.

* * *Stamens 4, the outer or lower pair longer, or only 2 with anthers, straight and nearly converging in pairs! Anthers 2-celled!*

  — *Corolla almost equally 4-lobed, quite small.*

2. **Mentha.** Calyx equally 5-toothed. Upper lobe of the corolla rather the broadest, and sometimes notched. Stamens 4, of equal length, not convergent. Flowers either in terminal spikes or in head-like whorled clusters, often forming interrupted spikes. Corolla purplish or whitish.

3. **Lycopus.** Calyx 4-5-toothed. Stamens 2, the upper pair, if any, without anthers. Flowers whorled, in dense axillary clusters.

  — — *Corolla more or less 2-lipped, but the lobes nearly equal in size; the tube not hairy inside, the only stamens with anthers 2.*

4. **Hedeoma.** Calyx 2-lipped, bulging at the lower side at the base, hairy in the throat. 2 stamens with good anthers, and 2 sterile filaments with false anthers. Low odorous plants, with bluish flowers in loose axillary clusters.

  — — — *Corolla 2-lipped, the lower of the 2 lobes much larger than the other 3; the tube with a beard-ring inside. Stamens 2 (or the lower only 2), much exserted.*

5. **Collinso'nia.** Calyx ovate, enlarged and turned down in fruit, 2-lipped. Corolla elongated, the lower lip toothed or fringed. Strong-scented plants with *yellowish flowers on slender pedicels* in terminal panicled racemes.

    + + + + *Corolla evidently 2-lipped. Stamens with anthers 4.*

6. **Saturc'in.** Calyx bell-shaped, not hairy in the throat, equally 5-toothed. Aromatic plants with narrow leaves and purplish spiked flowers.

* * * *Stamens only 2, parallel; the anthers only 1-celled! Corolla 2-lipped.*

7. **Monar'da.** Calyx tubular, nearly equally 5-toothed, hairy in the throat. Corolla elongated, strongly 2-lipped, the upper lip narrow. Stamens with long protruding filaments, each bearing a linear anther on its apex. Flowers large, in whorled heads surrounded by bracts.

* * * *Stamens 4, the upper or inner pair longer! Anthers approximate in pairs. Corolla 2-lipped.*

8. **Nep'eta.** Calyx obliquely 5-toothed. Anthers approaching each other in pairs under the inner lip of the corolla, the cells of each anther divergent. (See Figs. 57 and 58, Part I.)

9. **Calamin'tha.** Calyx tubular, *2-lipped*, often bulging below. Corolla 2-lipped, *the upper lip not arched*, the throat inflated. Flowers pale purple, in globular clusters which are crowded with awl-shaped hairy bracts.

* * * * *Stamens 4, the lower or outer pair longer! Anthers approximate in pairs. Corolla 2-lipped.*

10. **Brunel'a.** Calyx 2-lipped, flat on the upper side, closed in fruit; *the upper lip 3-toothed, the lower 2-cleft*. Filaments 2-toothed at the apex, the lower tooth bearing the anther. Flowers violet, in a close terminal spike or head which is very leafy-bracted.

11. **Scutella'ria.** Calyx 2-lipped, short, closed in fruit, the lips rounded and entire, *the upper with a projection on the back*. Corolla blue or violet, the tube elongated and somewhat curved. Anthers of the lower stamens 1-celled, of the upper 2-celled. Flowers solitary in the axils of the upper leaves, or in axillary or terminal 1-sided racemes.

12. **Marru'bium.** Calyx 10-toothed, the teeth spiny and recurved after flowering. Stamens 4, *included in the corolla-tube*. Whitish woolly plants with small white flowers in head-like whorls.

13. **Galeop'sis.** Calyx 5-toothed, the teeth spiny. The middle lobe of the lower lip of the corolla inversely heart-shaped, *the palate with 2 teeth at the sides*. Stamens 4, *the anthers opening cross-wise*. Flowers purplish, in axillary whorls.

14. **Stach'ys.** Calyx 5-toothed, beset with stiff hairs, the teeth spiny, diverging in fruit. Stamens 4, *the outer pair turned down after discharging their pollen*. Flowers purple, crowded in whorls, these at length forming an interrupted spike.

13. **Leonu'rus.** Calyx 5-toothed, the teeth spiny, and spreading when old. The middle lobe of the lower lip of the corolla inversely heart-shaped. Flowers pale purple, in close whorls in the axils of the *cut-lobed leaves*.

1. TEUCRIUM. GERMANDER.

T. Canadense. (AMERICAN GERMANDER. WOOD SAGE.) Stem 1-3 feet high, downy. Leaves ovate-lanceolate, serrate, short-petioled, hoary beneath. Flowers in a long spike.—Low grounds.

2. MENTHA. MINT.

1. **M. vir'idis.** (SPEARMINT.) *Flowers in a narrow terminal spike.* Leaves ovate-lanceolate, wrinkled-veiny, unequally serrate, sessile.—Wet places.

2. M. piperita. (PEPPERMINT.) *Flowers in loose interrupted spikes.* Leaves ovate or ovate-oblong, acute, *petioled*. Plant smooth.—Wet places.

3. M. **Canadensis.** (WILD MINT.) *Flowers in axillary whorled clusters*, the uppermost axils without flowers. Stem more or less hairy, with ovate or lanceolate toothed leaves on short petioles.—Shady wet places.

3. LYCOPUS. WATER HOREHOUND.

1. **L. Virgin'icus.** (BUGLE-WEED.) *Calyx-teeth 4, bluntish.* Stems obtusely 4-angled, 6-18 inches high, producing slender runners from the base. Leaves ovate-lanceolate, toothed.— Moist places.

2. L. Europæus, var. sinuatus. *Calyx-teeth 5, sharp-pointed.* Stems sharply 4-angled, 1-3 feet high. Leaves varying from cut-toothed to pinnatifid.—Wet places.

4. HEDEOMA. MOCK PENNYROYAL.

H. pulegioides. (AMERICAN PENNYROYAL.) Stem 5-8 inches high, branching, hairy. Leaves oblong-ovate, obscurely serrate. Whorls few-flowered. Plant with a pungent aromatic odour.—Open woods and fields.

5. COLLINSONIA. HORSE-BALM.

C. Canadensis. (RICH-WEED. STONE-ROOT.) Stem smooth or nearly so, 1-3 feet high. Leaves serrate, pointed, petioled, 3-6 inches long.—Rich woods.

6. SATURE'IA. SAVORY.

S. hortensis. (SUMMER SAVORY.) Stem pubescent. Clusters few-flowered.—Escaped from gardens in a few localities.

7. MONAR'DA. HORSE-MINT.

1. **M. did'yma.** (OSWEGO TEA.) *Corolla bright red*, very showy. The large outer bracts tinged with red.—Along shaded streams.

2. **M. fistulo'sa.** (WILD BERGAMOT.) *Corolla purplish.* The outer bracts somewhat purplish.—Dry and rocky banks and woods.

8. NEP'ETA. CAT-MINT.

N. Cata'ria. (CATNIP.) Flowers in cymose clusters. Stem erect, downy, branching. Leaves oblong, crenate, whitish beneath. Corolla dotted with purple.—Roadsides.

9. CALAMIN'THA. CALAMINTH.

C. Clinopo'dium. (BASIL.) Stem hairy, erect, 1-2 feet high. Leaves ovate, nearly entire, petioled.—Thickets and waste places.

10. BRUNEL'LA. SELF-HEAL.

B. vulga'ris. (COMMON HEAL-ALL.) A low plant with oblong-ovate petioled leaves. Clusters 3-flowered, the whole forming a close terminal elongated head.—Woods and fields everywhere.

11. SCUTELLA'RIA. SKULLCAP.

1. **S. galericula'ta.** Flowers blue, ¾ of an inch long, solitary in the axils of the upper leaves. Stem nearly smooth, 1-2 feet high.—Wet places.

2. **S. lateriflora.** Flowers blue, ⅓ of an inch long, in 1-sided racemes. Stem upright, much branched, 1-2 feet high.—Wet places.

12. MARRUBIUM. HOREHOUND.

M. vulgare. Leaves round-ovate, crenate-toothed. Calyx with 5 long and 5 short teeth, re-curved.—Escaped from gardens in some places.

13. GALEOPSIS. HEMP-NETTLE.

G. **Tetra'hit.** (COMMON HEMP-NETTLE.) Stem bristly-hairy, swollen **below** the **joints.** Leaves ovate, coarsely serrate. Corolla often with a **purple** spot on the lower **lip.**—Waste places and fields.

14. STACHYS. HEDGE-NETTLE.

S. palustris, var. aspera. Stem 2-3 feet high, 4-angled, the angles beset with stiff **reflexed** hairs or bristles.—Wet grounds.

15. LEONURUS. MOTHERWORT.

L. Cardi'aca. (COMMON MOTHERWORT.) Stem tall. Leaves long-petioled, the lower palmately **lobed,** the upper 3-cleft. Upper lip **of the corolla** bearded.—Near dwellings.

ORDER LVIII. BORRAGINA'CEÆ. (BORAGE FAMILY.)

Herbs, with a deeply **4-lobed ovary, forming 4 seed-like** nutlets, as in the last Order, but *the corolla is regularly 5-lobed with 5 stamens inserted upon its tube.*

Synopsis of the Genera.

* Corolla without any *scales in the throat.*

1. **Echium.** Corolla with a funnel-form tube and a spreading border of 5 somewhat unequal *lobes. Stamens* inserted, unequal. Flowers bright blue with a purplish tinge, in racemed clusters. Plant bristly.

* * Corolla with 5 scales somewhat *closing the throat.*

2. **Symphytum.** Corolla tubular-funnel-form, with short spreading lobes; scales subulate-awl. Flowers pale red or white, in nodding raceme-like clusters, the last written in pairs. Nettles small. Coarse hairy herbs.

3. **Echinospermum.** Nutlets prickly on the margins. Corolla salver-shaped, the lobes rounded; scales short and blunt. Flowers blue, small in leafy bracted racemes. Plant rough-hairy.

4. **Cynoglossum.** Nutlets prickly all over. Corolla funnel-form; scales blunt. Flowers red, purple or pale blue; the stamens which are naked above, but usually bearded at the base. Strong-scented coarse herbs.

* * * Corolla open, the scales or folds not sufficient **to completely close the** throat. Nutlets smooth.

5. **Onosmodium.** Corolla tubular, the **5 lobes acute** and erect or converging. Anthers more or less flaments very short. Style thread-like, much exserted. Flowers greenish or yellowish-white. Rather tall stout plants, shaggy with spreading bristly hairs.

6. **Lithosper'mum.** Corolla funnel-form or salver-shaped, the 5 lobes of the spreading limb rounded. Anthers almost sessile. Root mostly red. Flowers small and almost white, or large and deep yellow, scattered or spiked and leafy-bracted.

7. **Myoso'tis.** Corolla salver-shaped, with a very short tube, *the lobes convolute in the bud;* scales or appendages of the throat blunt and arching. Flowers blue, in racemes without bracts. Low plants, mostly in wet places.

1. E'CHIUM. Viper's Bugloss.

E. vulga're. (BLUE-WEED.) Stem erect, 2 feet high. Leaves sessile, linear-lanceolate. Flowers showy, in lateral clusters, the whole forming a long narrow raceme.—Roadsides; common in Eastern Ontario.

2. SYM'PHYTUM. Comfrey.

S. officina'le. (COMMON COMFREY.) Stem winged above by the decurrent bases of the leaves, branched. Leaves ovate-lanceolate or lanceolate.—Moist soil; escaped from gardens.

3. ECHINOSPER'MUM. Stickseed.

E. Lap'pula. A very common roadside weed, 1–2 feet high, branching above. Leaves lanceolate, rough. Nutlets warty on the back, with a double row of prickles on the margin.

4. CYNOGLOS'SUM. Hound's-Tongue.

1. **C. officina'le.** (COMMON HOUND'S-TONGUE.) *Flowers red-purple.* Upper leaves lanceolate, sessile. Stem soft-pubescent. Nutlets rather flat.—A common weed in fields and along roadsides.

2. **C. Virgin'icum.** (WILD COMFREY.) *Flowers pale blue.* Stem roughish with spreading hairs. Leaves few, lanceolate-oblong, *clasping.* Racemes corymbed, raised on a long naked peduncle.—Rich woods.

3. **C. Morisoni.** (BEGGAR'S LICE.) *Flowers pale blue or white.* Stem hairy, *leafy,* with broadly spreading branches. Leaves taper-pointed and tapering at the base. Racemes panicled, forking, widely spreading. Pedicels of the flowers reflexed in fruit.—Open woods and thickets.

5. ONOSMO'DIUM. FALSE GROMWELL.

O. Carolinia'num. Stem stout. 3–4 feet high. Leaves ovate-lanceolate, acute. *Lobes of the corolla very hairy outside.* —Banks of streams.

6. LITHOSPER'MUM. GROMWELL. PUCCOON.

1. **L. arven'se.** (CORN GROMWELL.) Stem 6–12 inches high, erect. Leaves lanceolate or linear. *Flowers almost white.* —Sandy banks.

2. L. **hirtum.** (HAIRY PUCCOON.) Stem 1–2 feet high. Stem-leaves lanceolate or linear; those of the flowering branches ovate-oblong, ciliate. *Corolla deep yellow, woolly at the base inside.* —Dry woods.

7. MYOSO'TIS. FORGET-ME-NOT.

M. palustris, var. **laxa.** (FORGET-ME-NOT.) Stem ascending from a creeping base, about a foot high, loosely branched. Corolla pale blue, with a yellow eye. Pedicels spreading.—Wet places.

ORDER LIX. HYDROPHYLLA'CEÆ. (WATERLEAF F.)

Herbs, with alternate cut-toothed or lobed leaves, and regular pentamerous and pentandrous flowers very much like those of the last Order, *but having a 1-celled ovary with the seeds on the walls (parietal).* Style 2-cleft. Flowers mostly in 1-sided cymes which uncoil from the apex. The only common Genus is   .

HYDROPHYLLUM. WATERLEAF.

H. Virgin'icum. Corolla bell-shaped, the 5 lobes convolute in the bud; the tube with 5 folds down the inside, one opposite each lobe. Stamens and style exserted, the filaments bearded below. Stem smoothish. **Leaves** pinnately cleft into 5–7 divisions, the latter ovate-lanceolate, pointed, cut-toothed. *Calyx-lobes very narrow, bristly-ciliate.* Flowers white or pale blue. Rootstocks scaly-toothed.—Moist woods.

ORDER LX. **POLEMONIA'CEÆ.** (POLEMONIUM F.)

Herbs, with regular pentamerous and pentandrous flowers, *but a 3-celled ovary and 3-lobed style.* *Lobes of the corolla convolute in the bud.* Calyx persistent. Represented commonly with us by only one Genus,

PHLOX. PHLOX.

P. divarica'ta. Corolla salver-shaped, with a long tube. Stamens short, *unequally inserted.* Stem ascending from a prostrate base, somewhat clammy. Leaves oblong-ovate. Flowers lilac or bluish, in a spreading loosely-flowered cyme. *Lobes of the corolla mostly obcordate.*—Moist rocky woods.

ORDER LXI. **CONVOLVULA'CEÆ.** (CONVOLVULUS F.)

Chiefly twining or trailing herbs, with alternate leaves and regular flowers. Sepals 5, imbricated. Corolla 5-plaited or 5-lobed and convolute in the bud. Stamens 5. Ovary 2-celled.

Synopsis of the Genera.

1. **Cal'ste'gia.** *Calyx enclosed in 2 large leafy bracts.* Corolla funnelform, the border obscurely lobed. Pod 4-seeded.
2. **Cus'cuta.** *Leafless parasitic slender twiners,* with yellowish or reddish stems, attaching themselves to the bark of other plants. Flowers small, mostly white, clustered. Corolla bell-shaped. *Stamens with a fringed appendage at their base.*

1. CALYSTE'GIA. BRACTED BINDWEED.

1. **C. se'pium.** (HEDGE BINDWEED.) *Stem mostly twining.* Leaves halbe d-shaped. Peduncles 4-angled. Corolla commonly rose-coloured.—Moist banks.

2. **C. spithamæ'a.** Stem low and simple, upright or ascending, *not twining,* 6-12 inches high. Leaves oblong, more or less heart-shaped at the base. Corolla white.—Dry soil.

2. CUS'CUTA. DODDER.

C. Grono'vii. St ms resembling coarse threads, spreading themselves over herbs and low bushes.—Low and moist shady places.

ORDER LXII. **SOLANA'CEÆ.** (NIGHTSHADE FAMILY.)

Rank-scented herbs (or one species shrubby), with colourless bitter juice, alternate leaves, and regular pentamerous and pentandrous flowers, *but a 2-celled ovary, with the placenta in the axis.* Fruit a many-seeded berry or pod.

Synopsis of the Genera.

1. **Sola'num.** Corolla wheel-shaped, 5-lobed, the margins turned inward in the bud. *Anthers coming* around the style, the cells opening by pores at the apex; filaments very short. The larger leaves often with an accompanying smaller one. Fruit a berry.

2. **Phy'salis.** Calyx 5-cleft, enlarging after flowering, becoming at length much inflated and bundled, and enclosing the berry. Corolla between wheel-shaped and funnel-form. *Anthers separate*, opening lengthwise.

3. **Hyoscy'amus.** *Fruit a pod, the top coming off like a lid.* Calyx urn-shaped, 5-lobed, persistent. Corolla funnel-form, *oblique*, the limb 5-lobed, dull coloured and veiny. Plant clammy-pubescent.

4. **Datu'ra.** Fruit a large prickly pod. Calyx long, 5-angled, not persistent. Corolla very large, funnel-form, strongly plaited in the bud, with 5 pointed lobes. Stigma 2-lipped. Rank-scented weeds, with the showy flowers in the forks of the branching stems.

1. SOLA'NUM. NIGHTSHADE.

1. **'S. Dulcama'ra.** (BITTERSWEET.) Stem somewhat shrubby and climbing. Leaves ovate and heart-shaped, *the upper halberd-shaped, or with two small lobes at the base.* Flowers violet-purple, in small cymes. *Berries red.*—Near dwellings, and in moist grounds.

2. **S. nigrum.** (COMMON NIGHTSHADE.) *Stem low and spreading, branched. Leaves ovate, wave-toothed.* Flowers small, white, drooping in umbel-like lateral clusters. *Berries black.*—Fields and damp grounds.

2. PHY'SALIS. GROUND CHERRY.

P. visco'sa. Corolla greenish-yellow, brownish in the centre. Anthers yellow. Leaves ovate or heart-shaped, mostly toothed. Berry orange, sticky.—Sandy soil.

3. HYOSCYAMUS. HENBANE.

H. niger. (BLACK HENBANE.) Escaped from gardens in some localities. Corolla dull yellowish, netted with purple veins. Leaves clasping, sinuate-toothed. A strong-s.ented and poison-ous herb.

4. DATURA. STRAMONIUM. THORN-APPLE.

D. Stramo'nium. (COMMON THORN-APPLE.) Stem green. Corolla white, 3 inches long. Leaves ovate, sinuate-toothed.— Roadsides.

ORDER LXIII. GENTIANA'CEÆ. (GENTIAN F.)

Smooth herbs, *distinguished by having a 1-celled ovary with seeds on the walls, either in 2 lines or on the whole inner surface.* Leaves mostly opposite, simple, and sessile, but in one Genus alternate and compound. Stamens as many as the lobes of the regular corolla and alternate with them. Stigmas 2. Calyx persistent. Juice colourless and bitter.

Synopsis of the Genera.

1. **Hale'nia.** Corolla 4-lobed, *the lobes all spurred at the base.* Flowers yellowish or purplish, somewhat cymose.

2. **Gentia'na.** Corolla not spurred, 4-5 lobed, mostly funnel-form or bell-shaped, generally with teeth or folds in the sinuses of the lobes. Stigmas 2, persistent. Pod oblong. Seeds innumerable. Flowers showy, in late summer and autumn.

3. **Bartoni'a.** Corolla short funnel-form, 5-lobed, seed at the base of the corolla tube. *Leaves alternate,* compressed, and reduced to scales. *Flowers in a raceme at the summit of a naked scape,* white or tinged with pink.

1. HALE'NIA. SPURRED GENTIAN.

H. deflexa. Stem erect, 9–18 inches high. Leaves 3–5 nerved, those at the base of the stem oblong-s.athulate, petioled; the upper acute and sessile or nearly so. Spurs of the corolla curved.—Not common in Ontario, but reported from Ancaster and Collingwood. Common on the Lower St. Lawrence.

2. GENTIA'NA. GENTIAN.

1. **G. crini'ta.** (FRINGED GENTIAN.) Corolla funnel-form, 4-lobed, *the lobes fringed on the margins*; no plaited folds in the sinuses. *Flowers sky-blue, solitary on long naked stalks* terminating the stem on simple branches. *(Grows biennial).* Leaves lance-shaped, or ovate-lanceolate. Low grounds.

2. **G. deton'sa** (SMALLER FRINGED G.) is distinguished from No. 1 by the shorter or almost inconspicuous fringe of the corolla, the linear or lance-linear leaves, and the broader ovary.— Moist grounds, chiefly in the Niagara District.

3. **G. alba.** (WHITISH G.) Corolla inflated-club-shaped, *at length open*, 5-lobed, the lobes about twice as long as the toothed *appendages in the sinuses.* Flowers *greenish-white* or *yellowish*, sessile, crowded in a terminal cluster. Anthers usually cohering. Leaves lance-ovate, with a clasping heart-shaped base.— Low grounds.

4. **G. Andrews'ii.** (CLOSED G.) Corolla inflated-club-shaped, *closed at the mouth*, the apparent lobes being really the large fringed-toothed appendages. *Flowers blue*, in a close sessile terminal cluster. Anthers cohering. Leaves ovate-lanceolate from a narrower base.—Low grounds; common northward, flowering later than No. 3.

3. MENYAN'THES. BUCKBEAN.

M. trifolia'ta. A common plant in bogs and wet places northward. The bases of the long petioles sheathe the lower part of the scape, or thick rootstock, from which they spring. Plant about a foot high.

ORDER LXIV. APOCYNACEÆ. (DOGBANE FAMILY.)

Herbs or slightly shrubby plants, with milky juice, opposite simple entire leaves, and regular pentamerous and pentandrous flowers with the lobes of the corolla convolute in the bud. *Distinguished by having 2 separate ovaries*, but the 2 stigmas united. Calyx free from the ovaries. Anthers converging round the stigmas. Seeds with a tuft of down on the apex. Represented with us only by the Genus

APO'CYNUM. DOGBANE.

1. **A. androsæmifo'lium.** (SPREADING DOGBANE.) Corolla bell-shaped, 5-cleft, *pale rose-co'oured, the lobes turned back. Branches of the stem widely forking.* Flowers in loose rather spreading cymes. Leaves ovate, petioled. Fruit 2 long and slender diverging pods.—Banks and thickets.

2. **A. cannab'inum.** (INDIAN HEMP.) Lobes of the *greenish-white* corolla not turned back. *Branches erect.* Cymes closer than in No. 1, and the flowers much smaller.—Along streams.

ORDER LXV. **ASCLEPIADA'CEÆ.** (MILKWEED F.)

Herbs with milky juice and opposite or whorled simple entire leaves. Pods, seeds, and anthers as in the last Order, *but the anthers are more closely connected with the stigma, the (reflexed) lobes of the corolla are valvate in the bud, the pollen is in waxy masses, and the (monadelphous) short filaments bear 5 curious hooded bodies behind the anthers.* Flowers in umbels. Commonly represented by only one Genus, which is typical of the whole Order

ASCLE'PIAS. MILKWEED.

1. **A. cornu'ti.** (COMMON MILKWEED.) Stem tall and stout. Leaves oval or oblong, short-petioled, pale green, 4–8 inches long. *Flowers dull greenish-purple. Pods ovate, soft spiny, woolly.*—Mostly in dry soil; very common.

2. **A. phytolaccoi'des.** (POKE-MILKWEED.) Stem tall and smooth. Leaves broadly ovate, acute at both ends, short-petioled. *Pedicels loose and nodding, very long and slender.* Corolla greenish *with the hooded appendages white.* Pods minutely downy, *but not warty.*—Moist thickets.

3. **A. incarna'ta.** (SWAMP M.) Stem tall, leafy, branching, and smooth. Leaves oblong-lanceolate, acute, obscurely heart-shaped at the base. *Flowers rose-purple. Pods very smooth and glabrous.*—Swamps and low grounds.

4. **A. tubero'sa.** (BUTTERFLY-WEED. PLEURISY-ROOT.) Stem very leafy, branching above, rough-hairy. Leaves linear or ob-

long-lanceolate, chiefly scattered. Corolla greenish-orange, *with the hoods bright orange-red*. Pods hoary.—Dry hill-sides and fields; almost destitute of milky juice.

ORDER LXVI. OLEA'CEÆ. (OLIVE FAMILY.)

The only common representative Genus of this Order in Canada is Fraxinus (Ash). The species of this Genus are trees with pinnate leaves, and polygamous or diœcious flowers without petals and mostly also without a calyx; stamens only 2, with large oblong anthers. Fruit a 1–2 seeded samara. Flowers insignificant, from the axils of the previous year's leaves.

FRAXINUS. ASH.

1. F. Americana. (WHITE ASH.) *Fruit winged from the apex only, the base cylindrical. Branchlets and petioles smooth and glabrous.* Calyx very minute, persistent. Leaflets 7–9, stalked.—Rich woods.

2. F. pubes'cens (RED ASH) has the branchlets and petioles softly-pubescent, and the fruit acute at the base, flattened, and gradually expanding into the long wing above.—Same localities as No. 1.

3. F. sambucifolia. (BLACK or WATER ASH.) Branchlets and petioles smooth. Leaflets 7–9, sessile, serrate. *Fruit winged all round.* Calyx wanting, and the flowers consequently naked. —Swamps.

DIVISION III. APETALOUS EXOGENS.

Flowers destitute of corolla and sometimes also of calyx.

ORDER LXVII. ARISTOLOCHIA'CEÆ. (BIRTHWORT F.)

Herbs with perfect flowers, the tube of the coloured calyx at length to the stipuled many-seeded ovary. Leaves heart-shaped or kidney-shaped, on long petioles from a thick rootstock. Stamens 12 or 6. Flowers solitary. Calyx dull-coloured, the lobes valvate in the bud.

AS'ARUM. WILD-GINGER.

A. Canadense. Radiating stigmas 6. Leaves only a single pair, kidney-shaped, and rather velvety, the peduncle in the fork between the petioles, close to the ground. Rootstock aromatic. Calyx brown-purple inside, the spreading lobes pointed. —Rich woods.

ORDER LXVIII. PHYTOLACCA'CEÆ. (POKEWEED F.)

Herbs with alternate leaves and perfect flowers, resembling in most respects the plants of the next Order, *but the ovary is composed of several carpels in a ring, forming a berry in fruit.* Only one Genus and one Species.

PHYTOLACCA. POKEWEED.

P. decan'dra. (COMMON POKE.) Calyx of 5 rounded white sepals. Ovary green, of 10 1-seeded carpels united into one. Styles 10, short and separate. Stamens 10. Fruit a crimson or purple 10-seeded berry. Stem very tall and stout, smooth. Flowers in long racemes opposite the leaves.—Rich soil.

ORDER LXIX. CHENOPODIA'CEÆ. (GOOSEFOOT F.)

Homely herbs, with more or less succulent leaves (chiefly alternate), and small greenish flowers mostly in interrupted spikes. Stamens usually as many as the lobes of the calyx and opposite them. Ovary 1-celled and 1-ovuled, forming an achene or utricle in fruit. Stigmas mostly 2.

Synopsis of the Genera.

1. **Chenopo'dium.** Weeds with (usually) mealy leaves, and very small greenish sessile flowers in small panicled spiked clusters. Calyx 5-cleft, more or less enveloping the fruit. Stamens mostly 5; filaments slender.

2. **Bli'tum.** Flowers in heads which form interrupted spikes. Calyx becoming fleshy and bright red in fruit, so that the clusters look something like strawberries. Leaves triangular and somewhat halberd-shaped, sinuate toothed.

3. **Salso'a,** with fleshy awl-shaped sharp-pointed leaves, is not uncommon on the Lower St. Lawrence and the sea-coast.

1. CHENOPO'DIUM. Goosefoot. Pigweed.

1. C. album. (Lamb's-Quarters.) Stem upright, 1-3 feet high. Leaves varying from rhombic-ovate to lanceolate, more or less toothed, *mealy, as are also the dense flower-clusters.*— Extremely common in cultivated soil.

2. C. hy'bridum. (Maple-leaved Goosefoot.) *Bright green.* Stem widely branching, 2-4 feet high. Leaves thin, large, triangular, heart-shaped, sinuate-angled, the angles extended into pointed teeth. Panicles loose, leafless. Plant with a rank unpleasant odour.—Waste places.

3. C. Bo'trys. (Jerusalem Oak.) Not mealy, but *sticky;* low, spreading, sweet-scented. Leaves deeply sinuate, slender-petioled. Racemes in divergent corymbs.—Roadsides; escaped from gardens.

2. BLI'TUM. Blite.

B. capita'tum. (Strawberry Blite.) Stem ascending, branching. Leaves smooth. The axillary head-like clusters very conspicuous in fruit.—Dry soil, margins of woods, &c.

Order LXX. AMARANTA'CEÆ. (Amaranth F.)

Homely weeds, a good deal like the plants of the last Order, but the flower-clusters are interspersed with *dry and chaff-like* (sometimes coloured) persistent bracts, *usually 3 to each flower.* Only one common Genus with us.

AMARANTUS. Amaranth.

1. A. panicula'tus. *Reddish flowers* in terminal and axillary slender spikes, the bract awn-pointed.—In the neighbourhood of gardens.

2. A. retroflex'us. (Pigweed.) *Flowers greenish,* in spikes forming a stiff panicle. Leaves a dull green, long-petioled, ovate, wavy-margined. *Stem erect.*—Common in cultivated soil.

3. A. albus. *Flowers greenish,* in small close axillary clusters. *Stem low and spreading.*—Roadsides.

ORDER LXXI. POLYGONACEÆ. (BUCKWHEAT.)

Herbs, *well marked by* the stipules of the alternate leaves *being in the form of membranous* sheaths above the usually *swollen joints of the stem.* Flowers usually perfect. Calyx 4–6 parted. Stamens 4–9, inserted on the base of the calyx. Stigmas 2 or 3. Ovary 1-celled, with a single ovule rising from the base, forming a little nutlet.

Synopsis of the Genera.

1. **Polyg'onum.** Sepals 5 (occasionally 4), often coloured and petal-like, *persis'tent*, embracing the 3-angled or sometimes flattish nutlet or achene. Flowers in racemes or spikes, or sometimes in the axils.

2. **Rumex.** Sepals 6, *the 3 outer ones herbaceous and* spreading in fruit, the 3 inner (called *valves*) somewhat petal-like and, after flowering, convergent over the 3-angled achene, *often with a* tubercle-like projection on *the back.* Stamens 6. Styles 3. Flowers usually in closed whorls, the latter in panicled racemes.

3. **Fagopy'rum.** Calyx 5-parted, petal-like. Stamens 8, with honey glands between them. Styles 3. Achene 3-angled. Flowers white, in panicles. Leaves triangular heart-shaped or halberd-shaped.

1. POLYG'ONUM. KNOTWEED.

* *Flowers along the stem,* greenish, greenish-white, nearly *sessile in the axils of the small* leaves. Sheaths cut-fringed *or torn.*

1. **P. avicula're.** (KNOTGRASS or BIRDGRASS.) A weed everywhere in yards and waste places. *Stem prostrate and spreading.* Stamens chiefly 5. Achene 3-sided. Stigmas 3. Leaves sessile.

* * *Flowers in terminal* spikes or heads, usually rose-coloured *or pinkish, densely crowded together.*

+ *Leaves not* heart-shaped at the base, on short stalks.

2. **P. incarna'tum.** *Sheaths* not fringed. Stem smooth, 3–6 feet high. Leaves long, tapering to a long narrow point, rough on the midrib and margins. Flowers rose-colour, *nodding.* Stamens 6. Styles 2. Achene flat or concave, in wet muddy places along streams and ponds.

3. P. Pennsylva'nicum. Sheaths not fringed. Stem 1-3 feet high, *the upper branches and the peduncles bristly with stalked glands.* Spikes thick, erect. *Stamens 8.* Achene flat.—Low open grounds.

4. **P. Persica'ria.** (LADY'S THUMB.) Sheaths with a somewhat ciliate border. *Stem nearly smooth*, a foot or more in height. *Leaves with a dark blotch on the middle* of the upper surface. Spikes dense, erect, on naked peduncles. Stamens 6. Achene flat or 3-angled, according as the stigmas are 2 or 3.— Very common near dwellings in moist ground.

5. P. amphib'ium. (WATER PERSICARIA.) Spike of flowers dense, oblong, dense, rose-red. *Stem floating* in shallow wat-r or rooting in soft mud. Leaves long-petioled, often floating. Sheaths not bristly-fringed. *Stamens 5.* Stigmas 2.—In shallow water, mostly northward.

6. P. hydropiperoi'des. (MILD WATER-PEPPER.) Stem slender, 1-3 feet high, *in shallow water.* Leaves narrow, roughish. *Sheaths hairy and fringed with long bristles.* Spikes slender, erect, pale rose-colored or whitish. Stamens 8. Stigmas 3. Achene 3-angled.—In shallow water.

7. **P. acre.** (WATER SMARTWEED.) Sheaths fringed with bristles. *Leaves transparent-dotted.* Stem rooting at the decumbent base, 2-4 feet high, *in shallow water or muddy soil.* Leaves narrow, taper-pointed. Spikes slender, erect, pale rose-colored. Stamens 8. Achene 3-angled.—Muddy soil or shallow water.

+ + *Leaves heart-shaped or sagittate. Sheaths much longer on one side than on the other.*

8. P. arifo'lium (HALBERD-LEAVED TEAR-THUMB) with grooved stem, halberd-shaped long-petioled leaves, flowers in short loose racemes, 6 stamens, and a flattish achene, is not uncommon on the Lower St. Lawrence.

9. P. sagitta'tum. (ARROW-LEAVED TEAR-THUMB.) Stem 4-angled, the angles beset with reflexed minute prickles, by which the plant is enabled to climb. Leaves arrow-shaped. Stamens 8. Achene 3-angled.—Common in low grounds.

10. **P. Convol'vulus.** (BLACK BINDWEED.) *Stem twining*, not prickly but *roughish*. Flowers in loose panicled racemes, 3 of the calyx-lobes ridged in fruit. Leaves heart-shaped and partly halberd-shaped. Not climbing so high as the next.—Cultivated grounds and waste places.

11. **P. dumeto'rum.** (CLIMBING FALSE BUCKWHEAT.) Stem twining high, *smooth;* 3 of the calyx-lobes *winged in fruit.*—Moist thickets.

2. RUMEX. DOCK. SORREL.

* *Herbage not sour, nor the leaves halberd-shaped.*

1. **R. orbicula'tus.** (GREAT WATER DOCK.) Growing in marshes. Stem erect, stout, 5-6 feet high. Leaves lanceolate, *not wavy-margined*, often over a foot long. Flowers nodding *on thread-like pedicels.* Valves nearly orbicular, finely net-veined, each with a grain on the back.—Wet places.

2. **R. salicifo'lius** (WHITE DOCK) may be looked for in marshes on the sea-coast and far northward. The whorls of flowers are dense and form a very conspicuous spike, owing to the great size of the grains on the back of the valves.

3. **R. crispus.** (CURLED DOCK.) *Leaves with strongly wavy or curly margins*, lanceolate. Whorls of flowers in long wand-like racemes. Valves grain-bearing.—Cultivated soil and waste places.

4. **R. obtusifo'lius.** (BITTER DOCK.) Lowest leaves oblong heart-shaped, obtuse, only slightly wavy-margined; the upper oblong-lanceolate, acute. Whorls loose, *distant.* Valves somewhat halberd-shaped, *sharply toothed at the base,* usually *one only* grain-bearing.—Waste grounds.

* * *Herbage sour; leaves halberd-shaped.*

5. **R. Acetosel'la.** (FIELD OR SHEEP SORREL.) Stem 6-12 inches high. *Flowers diœcious,* in a terminal naked panicle.—A very common weed in poor soil.

3. FAGOPY'RUM. BUCKWHEAT.

F. esculentum. (BUCKWHEAT.) Old fields and copses, remaining after cultivation.

ORDER LXXII. **LAURA'CEÆ.** (LAUREL FAMILY.)

Trees or shrubs with spicy-aromatic bark and leaves, the latter simple (often lobed), alternate, and marked with small transparent dots (visible under a lens). Sepals 6, petal-like. Flowers diœcious. Stamens in sterile flowers 9, inserted at the base of the calyx. Anthers opening by uplifting valves. Ovary in fertile flowers free from the calyx, 1-celled, with a single ovule hanging from the top of the cell. Style and stigma 1. Fruit a 1-seeded drupe.

SAS'SAFRAS. SASSAFRAS.

S. officina'le. A small or moderate-sized tree with yellowish or greenish-yellow twigs and ovate or 3-lobed entire leaves. Flowers greenish-yellow, in naked corymbs, appearing with the leaves in the axils of the latter. Drupe blue, on a reddish pedicel. The 9 stamens in 3 rows, the 3 inner each with a pair of yellow glands at the base of the filament.—Rich woods, in Southern and Western Ontario.

ORDER LXXIII. **THYMELEA'CEÆ.** (MEZEREUM F.)

Shrubs with tough leather-like bark and entire leaves. Flowers perfect. Calyx tubular, resembling a corolla, pale yellow. Stamens twice as many as the lobes of the calyx (in our species 8). Style thread-like. Stigma capitate. Ovary 1-celled, 1-ovuled, free from the calyx. Fruit a berry-like drupe. Only one species in Canada.

DIRCA. LEATHERWOOD. MOOSE-WOOD.

D. palustris. A branching shrub, 2-5 feet high, with curious jointed branchlets and nearly oval leaves on short petioles. Flowers in clusters of 3 or 4, preceding the leaves. Filaments exserted, half of them longer than the others.—Damp woods.

ORDER LXXIV. **ELÆAGNA'CEÆ.** (OLEASTER F.)

Shrubs with diœcious flowers, and leaves which are scurfy on the under surface. The calyx-tube in the fertile flowers becomes fleshy and encloses the ovary, forming a berry-like fruit.

Otherwise the plants of this Order are not greatly different from those of the last.

SHEPHERD'IA. Shepherdia.

S. Canadensis. Calyx in sterile flowers 4-parted. Stamens 8. Calyx in fertile flowers urn-shaped, 4-parted. Berries yellow. Branchlets brown-scurfy. Leaves opposite, entire, ovate, green above, silvery-scurfy beneath, the small flowers in their axils.—Gravelly banks of streams and lakes.

Order LXXV. SANTALA'CEÆ. (Sandalwood F.)

Low herbaceous or partly woody plants (with us) with perfect flowers, these *greenish-white*, in terminal or axillary corymbose clusters. Calyx bell-shaped or urn-shaped, 4–5 cleft, adherent to the 1-celled ovary, lined with a 5-lobed disk, the stamens on the edge of the latter between its lobes and *opposite the lobes of the calyx, to which the anthers are attached by a tuft of fine hairs.* Fruit nut-like, crowned with the persistent calyx-lobes.

COMANDRA. Bastard Toad-flax.

C. umbella'ta. Stem 8–10 inches high, leafy. Leaves oblong, pale green, an inch long. Flower-clusters at the summit of the stem. Calyx-tube prolonged and forming a neck to the fruit. Style slender.—Dry soil.

Order LXXVI. EUPHORBIA'CEÆ. (Spurge F.)

Plants with milky juice and monœcious flowers, represented in Canada chiefly by the Genus

EUPHOR'BIA. Spurge.

Flowers monœcious, the sterile and fertile ones both destitute of calyx and corolla, *but both contained in the same 4–5 lobed cup-shaped involucre which resembles a calyx,* and therefore the whole will probably at first sight be taken for a single flower. Sterile flowers numerous, *each of a single naked stamen from the axil of a minute bract.* Fertile flower only 1 in each involucre; ovary 3-lobed, *soon protruded on a long pedicel;* styles 3, each 2-cleft. Peduncles terminal, often umbellate.

* *Leaves all similar and opposite, short-y lobed, green or blotched with brown above, furnished with scale-like or fringed stipules. Stems spreading or prostrate, much forked. Involucres in terminal or lateral clusters, or one involucre in each fork, the involucre invariably with 4 (mostly petal-like) glands in the sinuses.*

1. E. polygonifolia. Leaves *entire*, oblong-linear, mucronate, *very smooth. Stipules bristly*-fringed. *Peduncles* in the forks. *Glands of the involucre very small, not petal-like.* Pods obtusely angled —shores of the Great Lakes, in sandy or gravelly places.

2. **E.** maculata. *Leaves serrulate*, oblong-linear, *somewhat pubescent, with a brownish blotch in the centre*, very oblique at the base. *Peduncles in dense lateral clusters.* Glands of the involucre *with reddish petal-like attachments. Pods sharply angled.*—Roadsides.

3. E. hypericifolia. Stem ascending. Leaves *serrate*, often with *a red spot or with* red *margins*, oblique at the base, ovate-oblong or *oblong-linear. Peduncles in cymes at the ends of the branches. Glands* of the *involucre with white* or occasionally reddish petal-like *attachments.* Pod *smooth*, obtusely angled.— Cultivated soil.

* * *Involucres chiefly in terminal umbels, and their glands always without petal-like attachments. Leaves without stipules or blotches, those of the stem alternate or scattered, the floral ones usually of a different shape, and whorled or opposite.*

4. **E.** platyphylla. *Umbel 5-rayed.* Stem erect, 8-18 inches high. *Upper stem-leaves lanceolate*-oblong, acute, serrulate, the *uppermost heart-shaped*, the floral ones *triangular*-ovate and *cordate. Pod warty.* Shores of the Great Lakes.

5. E. Helioscopia. *Umbel first 5-rayed*, then *with 3*, and finally *merely forked. Stem ascending*, 6-12 inches high. Leaves all *obovate, rounded* or *notched at the* apex, serrate. *Pods smooth.—Along the Great Lakes.*

6. **E.** Cyparissias, *with densely* clustered stems, and crowded linear *stem-leaves (the floral ones round heart*-shaped), and a *many-rayed umbel, has escaped from gardens in some localities.*

ORDER LXXVII. **URTICA'CEÆ.** (NETTLE F.)

Herbs, shrubs, or trees, with monœcious or diœcious (or, in the Elms, sometimes perfect) flowers, with a regular calyx free from the 1–2 celled ovary which becomes a 1-seeded fruit. Stamens opposite the lobes of the calyx. This Order is divided into four well-marked Suborders, three of which are represented in Canada.

SUBORDER I. **ULMA'CEÆ.** (ELM FAMILY.)

Trees, with alternate simple leaves, and deciduous small stipules. Flowers often perfect. Styles 2. *Fruit a samara winged all round.*

1. **Ulmus.** Flowers in lateral clusters, earlier than the leaves, purplish or greenish-yellow. Calyx bell shaped, 4–9 cleft. Stamens 4–9; the filaments long and slender. Ovary 2-celled, but the samara only 1-seeded. Stigmas 2.

SUBORDER II. **URTI'CEÆ.** (NETTLE FAMILY.)

Herbs with watery juice and opposite or alternate leaves, often beset with stinging hairs. Flowers monœcious or diœcious, in spikes or racemes. Stamens as many as the sepals. Style only 1. Ovary 1-celled. Fruit an achene.

2. **Urti'ca.** *Leaves opposite.* Plant beset with stinging hairs. Sepals 4 in both sterile and fertile flowers. Stamens 4. Stigma a small sessile tuft. Achene flat, enclosed between the 2 larger sepals. Flowers greenish.

3. **Laport'ea.** *Leaves alternate.* Plant beset with stinging hairs. Sepals 5 in the sterile flowers, 4 in the fertile, 2 of them much smaller than the other 2. Stigma awl-shaped. Achene flat, *very oblique, reflexed on its winged pedicel.*

4. **Pi'lea.** Leaves opposite. *Whole plant very smooth and semi-transparent.* Sepals and stamens 3–4. Stigma a sessile tuft.

5. **Behme'ria.** Leaves mostly opposite. No stinging hairs. Sepals and stamens 4 in the sterile flowers. Calyx tubular or urn-shaped in the fertile ones, and enclosing the achene. Stigma long and thread like.

SUBORDER IV. **CANNABINEÆ.** (HEMP FAMILY.)

Rough herbs with watery juice and tough bark. Leaves opposite and *palmately compound.* Flowers diœcious. Sterile ones in compound racemes; stamens 5; sepals 5. Fertile ones in crowded clusters; sepal only 1, embracing the achene. Stigmas 2.

6. **Cannabis.** A rather tall rough plant with palmately compound leaves of 5-7 linear-lanceolate serrate leaflets.

1. ULMUS. Elm.

1. **U. fulva.** (RED OR SLIPPERY ELM.) Flowers nearly sessile. *Leaves very rough above, taper-pointed. Buds downy with rusty hairs.* A medium-sized tree, with mucilaginous inner bark.

2. **U. America'na.** (AMERICAN OR WHITE ELM.) *Leaves not rough above, abruptly pointed.* Flowers on drooping pedicels. *Buds glabrous.* A large ornamental tree, with drooping branchlets.—Moist woods.

2. URTICA. NETTLE.

U. gra'cilis. Stem slender, 2-6 feet high. Leaves ovate-lanceolate, pointed, serrate, 3-5 nerved from the base, nearly smooth, the long petioles with a few bristles. Flower-clusters in slender spikes.—Moist ground and along fences.

3. LAPORTEA. WOOD-NETTLE.

L. Canadensis. Stem 2-3 feet high. Leaves large, ovate, long-petioled, *a single 2-cleft stipule in the axil.*—Moist woods.

4. PILEA. RICHWEED. CLEARWEED.

P. pu'mila. Stem 3-18 inches high. Leaves ovate, coarsely toothed, 3-ribbed.—Cool moist places.

5. BOEHMERIA. FALSE NETTLE.

B. cylin'drica. Stem 1-3 feet high, smoothish. Leaves ovate-oblong or ovate-lanceolate, serrate, 3-nerved, long-petioled. Stipules separate.—Moist shady places.

6. CANNABIS. HEMP.

C. sati'va. (HEMP.) Common everywhere along roadsides and in waste places.

ORDER LXXVIII. PLATANA'CEÆ. (PLANE-TREE F.)

Represented only by the Genus

PLATANUS. PLANE-TREE. BUTTONWOOD.

P. occidenta'lis. (AMERICAN PLANE-TREE OR SYCAMORE.) A fine large tree found in Southwestern Ontario. Leaves alternate, rather scurfy when young, palmately lobed or angled, the lobes sharp-pointed; *stipules sheathing.* Flowers monœcious, both sterile and fertile ones in catkin-like heads, without calyx or corolla, but with small scales intermixed. Ovaries in the fertile flowers club-shaped, tipped with the thread-like simple style, and downy at the base. Fertile heads solitary, on slender peduncles. The white bark separates into thin plates.

ORDER LXXIX. JUGLANDA'CEÆ. (WALNUT F.)

Trees with alternate pinnate leaves and no stipules. Flowers monœcious. Sterile flowers in catkins. Fertile flowers solitary or in small clusters, with a regular 3–5 lobed calyx adherent to the ovary. Fruit a sort of drupe, the fleshy outer layer at length becoming dry and forming a husk, the inner layer hard and bony and forming the nut-shell. Seed solitary in the fruit, very large and 4-lobed. This Order comprises the Walnuts, Butternuts, and Hickories.

Synopsis of the Genera.

1. **Juglans.** Sterile flowers in *solitary* catkins from the previous year's wood. Filaments of the numerous stamens very short. Fertile flowers on peduncles at the ends of the branches. Calyx 4-to the 1, *with 4 small petals at the sinuses.* Styles and stigmas the latter fringed. *Exocarp or husk by-to without splitting.* Shell o the nut *very rough and irregu-larly furrowed.*

2. **Car'ya.** Sterile flowers in slender *clustered* catkins. Stamens 4-10, on very short filaments. Fertile flowers in small clusters at the ends of the branches. Calyx 4-cleft; no petals. Stigmas 4 or 1 large. *Exocarp 4-valved, drying and splitting away from the very smooth and bony nut-shell.*

1. JUGLANS. WALNUT.

1. **J. ciner'ea.** (BUTTERNUT.) Leaflets oblong-lanceolate, pointed, serrate. *Petioles and branchlets clammy.* Fruit oblong, *clammy.*—Rich woods.

2. **J. nigra.** (BLACK WALNUT.) Leaflets ovate-lanceolate, taper-pointed, serrate. Petioles downy but *not clammy.* Fruit spherical. Wood a darker brown than in the Butternut.—Rich woods; rare northward.

2. CARYA. Hickory.

1. C. alba. (Shell-bark Hickory.) *Leaflets 5, the lower pair* **much** smaller than the others. Husk **of the fruit** splitting completely into 4 valves. Nut thin, heart-shaped, **m**ucronate. Bark of the trunk rough, scaling off in rough **strips.—Rich** woods.

2. C. amara. (Swamp Hickory or **Bitternut.)** *Leaflets 7–11.* Husk of the fruit splitting half-way **down. Nut spherical, short-pointed.** Bark smooth, **not** scaling off.—Moist **ground.**

Order LXXX. CUPULIFERÆ. (Oak Family).

Shrubs or trees, with alternate **simple leaves, deciduous** stipules and monœcious flowers. **Sterile flowers in catkins** (but in Beech **in** small heads) ; **the fertile ones solitary or** clustered, and furnished with an **involucre which forms a** scaly cup or **a bur** surrounding **the nut.**

Synopsis of the **Genera.**

1. Quercus. Sterile flowers with a calyx including five **or several stamens** with slender filaments. Fertile flowers scattered or somewhat clustered, each in a cup-like involucre or cup, &c. Nut (the dry rounded **the base enclosed by the cup**) &c.

2. Castanea. Sterile flowers **in long slender catkins.** Calyx **6-parted.** Fertile flowers mostly 3 in each **involucre.** &c. prickly **forming** a bur of large spines. Stigmas linear, shaped in the seed, &c. (mostly 2 aborting them), who produce a dotted one above, but where there are indefinite one.

3. Fagus. Sterile flowers in a small head on drooping peduncles, their bell-shaped, 5-parted &c. &c. &c. the involucre, &c. consists of awl-shaped bracts to more together of the lower. Calyx five-toothed. Nut 3-angled, generally 2 together in the bur &c. soft of spinto. Bark very smooth and light grey.

4. Corylus. Sterile flowers in drooping catkins. &c. &c. a scale-like bracts and several small &c. &c. &c. &c. &c. the flowers in an obovoid bud, with a few stigmas. &c. &c. &c. &c. &c. stigmas and projecting but a &c. &c. &c. &c. in fruit enlarged and cut &c. &c. &c. &c. &c. &c. &c. leaves-budded and fruit of same year. Sterile flowers form on &c.

5. **Os'trya.** Sterile flowers in drooping catkins. *Calyx wanting.* Stamens several under each bract, *but not accompanied by bractlets.* Fertile flowers in short catkins, 2 under each bract, each ovary tipped with 2 long stigmas, and *surrounded by a tubular bractlet which, in fruit, becomes a greenish-white inflated bag, having the small nut in the bottom.*

6. **Carpi'nus.** Sterile flowers in drooping catkins. *Calyx wanting.* Stamens several under each bract; *no bractlets.* Fertile flowers much as in Ostrya, *but the bractlets surrounding the ovaries are not tubular but open, and in fruit become leaf-like, one on each side of the small nut.*

1. QUERCUS. OAK.

* *Acorns ripening the first year, and therefore borne on the new shoots. Lobes or teeth of the leaves not bristle-pointed.*

1. **Q. alba.** (WHITE OAK.) A large tree. Leaves (when mature) smooth, bright green above, whitish beneath, obliquely cut into few or several oblong entire lobes. The oblong nut much larger than the saucer-shaped rough cupule.—Rich woods.

2. **Q. macrocar'pa.** (BUR-OAK. MOSSY-CUP WHITE-OAK.) A medium-sized tree. Leaves deeply lobed, smooth above, pale or downy beneath. Acorn broadly ovoid, *half or altogether covered by the deep cup, the upper scales of which taper into bristly points making a fringed border.* Cup varying greatly in size, often very large.—Rich soil.

3. **Q. bi'color.** (SWAMP WHITE OAK.) A tall tree. Leaves sinuate-toothed, but hardly lobed, wedge-shaped at the base, downy or hoary beneath. Cup nearly hemispherical, about half as long as the oblong-ovoid acorn. sometimes with a fringed border. Peduncle in fruit larger than the petiole.—Low grounds.

* * *Acorns ripening the second year, and therefore borne on the previous year's wood, below the leaves of the season. Lobes or teeth of the leaves bristle-pointed.*

4. **Q. coccin'ea,** var. **tincto'ria.** (QUERCITRON, YELLOW-BARKED, OR BLACK OAK.) Leaves pinnatifid, slender-petioled, rather rounded at the base, rusty-downy when young, smooth and shining above when mature, often slightly pubescent beneath, turning brownish, orange, or dull red in the autumn. *Cup hemispherical with a more or less conical base,* covering half

or more of the rather small acorn.—Mostly in dry soil, but occasionally in moist places. Inner bark yellow, used for dyeing.

5. **Q. rubra.** (RED OAK.) A large tree. Leaves pinnatifid, turning dark red in the autumn. Cup saucer-shaped, sessile or nearly so, very much shorter than the oblong-ovoid acorn.—Rich and poor soil.

2. CASTA'NEA. CHESTNUT.

C. vesca, var. **America'na.** (CHESTNUT.) A large tree. Leaves oblong-lanceolate, pointed, coarsely and sharply serrate, acute at the base. Nuts 2 or 3 in each bur.—Southwestern Ontario.

3. FAGUS. BEECH.

F. ferrugin'ea. (AMERICAN BEECH.) A very common tree in rich woods, the branches horizontal. Leaves oblong-ovate, taper-pointed, toothed, the very straight veins terminating in the teeth.

4. COR'YLUS. HAZEL-NUT. FILBERT.

1. **C.** America'na. (WILD HAZEL-NUT.) Leaves roundish heart-shaped. Involucre spreading out above, leaf-like and cut-toothed.—Chiefly in Southwestern Ontario; in thickets.

2. **C.** rostra'ta. (BEAKED HAZEL-NUT.) A rather common shrub, easily distinguished from No. 1 by the involucre, which is prolonged into a narrow tube much beyond the nut, and is densely bristly-hairy.

5. OSTRYA. HOP-HORNBEAM. IRON-WOOD.

O. Virgin'ica. (IRON-WOOD.) A slender tree with brownish furrowed bark. Leaves oblong-ovate, taper-pointed, sharply doubly serrate. Fertile catkin like a hop in appearance. Wood very hard and close.—Rich woods.

6. CARPINUS. HORNBEAM.

C. America'na. (BLUE OR WATER BEECH.) Small trees with furrowed trunks and close smooth gray bark. Leaves ovate-oblong, pointed, doubly serrate.—Along streams. Resembling a Beech in general aspect, but with inflorescence like that of Iron-wood.

ORDER LXXXI. MYRICA'CEÆ. (SWEET-GALE F.)

Shrubs with monœcious or diœcious flowers, both sterile and fertile ones collected in short catkins or heads. *Leaves with resinous dots, usually fragrant.* Fruit a 1-seeded dry drupe or little nut, usually coated with waxy grains.

Synopsis of the Genera.

1. **Myri'ca.** Flowers chiefly diœcious, catkins lateral, each bract with a pair of bractlets underneath. Stamens in the sterile flowers 2–8. Ovary solitary in the fertile flowers, 1-celled, tipped with 2 thread-like stigmas, and surrounded by 2–4 small scales at the base. In our species the 2 scales form wings at the base of the nut.—A shrub, 3–5 feet high.

2. **Compto'nia.** A low shrub, a foot or more in height, with fern-like very sweet-scented leaves. Flowers monœcious. Sterile catkins cylindrical. Fertile ones spherical, the ovary surrounded by 8 awl-shaped persistent scales, so that the catkin resembles a bur.

1. MYRICA. BAYBERRY, WAX-MYRTLE.

M. Gale. (SWEET GALE.) Leaves wedge-lanceolate, serrate towards the apex, pale. The small nuts in crowded heads, and winged by the 2 scales.—Bogs.

2. COMPTO'NIA. SWEET-FERN.

C. asplenifo'lia. Leaves linear-lanceolate in outline, deeply pinnatifid, the lobes numerous and rounded.—Dry soil; observed at High Park, Toronto, and on the Pine barrens west of Angus, Ont.

ORDER LXXXII. BETULA'CEÆ. (BIRCH FAMILY.)

Trees or shrubs with monœcious flowers, both sorts in catkins, 2 or 3 ree flowers under each scale or bract of the catkin. Ovary 2-celled and 2-ovuled, but in fruit only 1-celled and 1-seeded. Fruit a small nut. Stigmas 2, long and slender. Twigs and leaves often aromatic.

Synopsis of the Genera.

1. **Bet'ula.** Sterile catkins long and pendulous, formed during summer and expanding the following spring ; each flower consisting of one small scale to which is attached 4 short filaments; 3 flowers under each scale of the catkin. Fertile catkins stout, oblong, the scales or bracts 3-lobed and with 2 or 3 flowers under each ; each flower a naked ovary, becoming a winged nutlet in fruit. Bark easily coming off in sheets.

2. Alnus. Catkins much as in Betula, but each fertile and sterile flower has a distinct 3-5 parted calyx. Catkins solitary or clustered at the ends of leafless branchlets or peduncles. Nuts wing.ess or nearly so.

1. BETULA. BIRCH.

1. B. lenta. (CHERRY BIRCH. SWEET OR BLACK BIRCH.) Bark of the trunk dark brown, close, aromatic; that of the twigs bronze-coloured. Wood rose-coloured. Leaves ovate, with somewhat heart-shaped base, doubly serrate, pointed, short-petioled. Fruiting catkins sessile, thick, oblong-cylindrical.—Moist woods.

2. B. lu'tea. (YELLOW OR GRAY BIRCH.) Bark of the trunk yellowish-gray, somewhat shreddy, scaling off in thin layers. Leaves hardly at all heart-shaped. Fruiting catkins thicker and shorter than in No. 1.—Moist woods.

3. B. papyra'cea. (PAPER OR CANOE BIRCH.) Bark of the trunk white, easily separating in sheets. Leaves ovate, taper-pointed, heart-shaped, long-petioled. Fruiting catkins cylindrical, usually hanging on slender peduncles.—Woods.

2. ALNUS. ALDER.

A. inca'na. (SPECKLED OR HOARY ALDER.) A shrub or small tree, growing in thickets in low ground along streams. Leaves oval or ovate, rounded at the base, serrate, whitish beneath. Flowers preceding the leaves in early spring.

ORDER LXXXIII. SALICA'CEÆ. (WILLOW F.)

Trees or shrubs with dioecious flowers, both sorts in catkins, one under each scale of the catkin. No calyx. Fruit 1-celled, many-seeded, the seeds furnished with tufts of down. (See Part I., section 36, for description of typical flowers.) This Order comprises the Willows and Poplars.

Synopsis of the Genera.

1. Salix. Trees with mostly long and pointed leaves and slender branches. Bracts or scales of the catkins even-toothed. Stamens mostly 2, sometimes more, but never so many as in No. 2. Stigmas short. Catkins appearing before or with the leaves.

2. Populus. Trees with broad and more or less heart-shaped leaves. Bracts of the catkins cut or torn at the apex. Stamens 12 to numerous, under each scale. Stigmas large. Catkins long and drooping, preceding the leaves.

1. SALIX.   WILLOW.

* *Catkins lateral and sessile, appearing before the leaves.   Scales dark red or brown, persistent.   No leaf-like bracts at the base of the catkins.   Stamens 2.*

+ *Leaves veiny, hairy or woolly, and with somewhat revolute margins.*

1. **S. can'dida.** (HOARY WILLOW.)   A shrub, not more than 3 or 4 feet high, growing in bogs and wet places; the twigs and leaves clothed with a web-like wool, giving the whole plant a whitish aspect.   Leaves lanceolate, narrow.   Stipules small, lanceolate, toothed.   Catkins cylindrical.

2. **S. hu'milis.** (PRAIRIE WILLOW.)   A shrub, 3-8 feet high, growing usually in dry or barren places.   Leaves lanceolate, not so taper-pointed as in No. 1, slightly downy above, thickly so beneath.   Stipules semi-ovate or moon-shaped with a few teeth, shorter than the petioles   Catkins ovoid.

+ + *Leaves smooth and shining above, not woolly beneath.   Catkins large, clothed with long glossy hairs.*

3. **S. dis'color.** (GLAUCOUS WILLOW.)   A shrub or small tree, 8-15 feet high, growing in low grounds and along streams.   Leaves lanceolate or ovate-lanceolate, irregularly toothed *in the middle of the margin,* entire at each end, white-glaucous beneath.   Stipules moon-shaped, toothed.

The 3 species just described frequently have compact heads of leaves, resembling cones, at the ends of the branches   This is probably a diseased condition due to puncturing by insects.

* * *Catkins lateral, preceding (or sometimes accompanying) the leaves.   Scales dark red or brown, persistent, 4 or 5 leaf-like bracts at the base of the catkin.*

4. **S. corda'ta.** (HEART-LEAVED WILLOW.)   A shrub or small tree, growing in wet grounds.   Leaves lanceolate, not always heart-shaped, sharply serrate, smooth.   Catkins cylindrical, leafy-bracted at the base.

* * * *Catkins lateral, appearing along with the leaves, leafy-bracted at the base. Stamens 2. Scales persistent.*

5. S liv'ida, var. **occidentalis.** (LIVID WILLOW.) A good-sized shrub, chiefly in moist situations. *Leaves oblong* or obo-vate-lanceolate, barely toothed, downy above, very veiny, hairy and glaucous beneath. Stipules semi-lunar, toothed. Ovary at length raised on a very slender stalk.

* * * * *Catkins long and loose, peduncled, not lateral but borne on the ends of the new shoots. Scales greenish-yellow, deciduous. Filaments hairy below.*

+ *Stamens 3-6 or more.*

6. S. lu'cida. (SHINING WILLOW.) A shrub or small bushy tree, growing along streams. *Leaves ovate-oblong* or *narrower*, with a long tapering point, shining on *both sides*, serrate. Sta-mens mostly 5.

7. S. nigra. (BLACK WILLOW.) A larger tree than No. 6, with a roughish black bark, growing along streams. *Leaves narrowly lanceolate*, tapering at each end, serrate, smooth, green on both sides. Stamens 3-6.

+ + *Stamens 2.*

8. S. longifolia. (LONG-LEAVED WILLOW.) A shrub or small tree, varying greatly in size, growing along streams in sandy or gravelly places. Leaves linear-lanceolate, very long, tapering to-wards both ends, nearly sessile, serrate with a few spreading teeth, grayish-hairy when young.

2. POP'ULUS. Poplar.

1. **P. tremuloides.** (AMERICAN ASPEN.) A tree with green-ish-white bark, and *roundish heart-shaped* leaves continually in a state of agitation, due to the lateral compression of the petiole, and the consequent susceptibility of the leaf to the least motion of the air. *Teeth of the leaves small.*

2. **P.** grandidentata (LARGE-TOOTHED ASPEN) has *roundish ovate* leaves with large irregular serrate teeth.

3. P balsamifera. (BALSAM POPLAR.) A tall tree, growing in swamps and along streams; the large buds varnished with resin-ous matter. Leaves ovate, tapering, finely serrate, whitish beneath. Stamens very numerous.

Subclass II. GYMNOSPERMS.

Ovules and seeds naked (not enclosed in a pericarp), and fertilized by the direct application of the
pollen. Represented in Canada by a single Order.

Order LXXXIV. CONIFERÆ. (Pine Family.)

Trees or shrubs with resinous juice and mostly monœcious
flowers, these in catkins except in the last genus (Taxus), in
which the fertile flower is solitary and the fruit berry-like.
Leaves awl-shaped or needle-shaped.—The Order comprises
three well-marked Suborders.

Suborder I. ABIETINEÆ. (Pine Family Proper.)

Fruit a true *cone*, the imbricated scales *in the axils of bracts*.
Ovules 2 on the inside of each scale at the base, in fruit coming off with a wing attached to each. (Plate I., Figs. 155, 156.)

* Cones not ripening till the second year.

1. **Pinus.** Leaves needle-shaped in fascicles in the axil of a
thin scale. Sterile stamens in spikes at the base, each consisting of many almost sessile anthers spirally arranged on an axis. Cone
more or less woody, the scales persistent, separating when ripe. Cotyledons
of the embryo several.

* Cones ripening the first year.

2. **Abies.** Leaves flattened or four-sided, often of two kinds on the new
shoots, evergreen. Scales falling when ripe, leaving the persistent axis
with thin scales.

3. **Larix.** Leaves needle-shaped, in fascicles or crowded on knobs of last
year's wood, many in each, deciduous in autumn; on the new shoots scattered, but few in each fascicle.

Suborder II. CUPRESSINEÆ. (Cypress Family.)

Fertile flowers of only a few scales, these set in the axils of
bracts, forming in fruit either a very small and berry-like dry cone,
or a sort of false berry owing to the thickening of the scales.

* Fruit a small, dry cone, with few scales.

4. **Thuja.** Leaves scale-like, imbricated in four rows, closely adherent to
the *flat branches*.

* * *Flowers mostly two two. Fruit berrylike,* black with a bloom.

5. Junip'erus. Leaves awlshaped or scale-like, sometimes of both shapes, evergreen, prickly-pointed, glaucous-coated on the upper surface, and (in our common species) in whorls of 3.

SUBORDER III. TAXINEÆ. (YEW FAMILY.)

Fertile flower solitary, consisting of a naked ovule surrounded by a disk which becomes pulpy and berry-like in fruit, enclosing the nut-like seed. *Berry red.*

6. Taxus. Flowers chiefly diœcious. Leaves evergreen, *mucronate,* rigid, scattered.—A low straggling bush, usually in the shade of other evergreens.

1. PINUS. PINE.

1. P. resino'sa. (RED PINE.) *Leaves in twos,* slender. Bark rather smooth, *reddish.*—Common northward.

2. P. strobus. (WHITE PINE.) *Leaves in fives,* slender. Bark smooth except on old trees, not reddish.—Common.

2. ABIES. SPRUCE. FIR.

1. A. nigra. (BLACK SPRUCE.) *Leaves needle-shaped and 4-sided, pointing in all directions.* Cones hanging, persistent, scales with thin edges.—Swamps and cold woods.

2. A. alba. (WHITE SPRUCE.) Leaves as in No. 1. Cones hanging, *deciduous,* the scales with thickish edges.—Swamps and cold woods.

3. A. Canadensis. (HEMLOCK SPRUCE.) *Leaves flat,* lighter beneath, *pointing only in two directions,* i.e. right and left on each side of the branch, distichous. Cones hanging, persistent.—Hilly or rocky woods.

4. A. balsamea. (BALSAM FIR.) *Leaves flat,* the lower surface whitish and the middle prominent, crowded, pointing mostly right and left on the branches. Cones erect on the upper sides of the branches, violet-coloured, *the scales sooner painted,*—swampy woods and swamps.

3. LARIX. LARCH.

L. America'na. (AMERICAN LARCH. TAMARACK.) A slender and very graceful tree with soft leaves in fascicles, falling off in autumn.—Swamps.

4. THUJA. ARBOR VITÆ.

T. occidentalis. (AMERICAN ARBOR VITÆ.) The well-known cedar of cedar-swamps.—Common.

5. JUNIP'ERUS. JUNIPER.

J. commu'nis. (COMMON JUNIPER.) A spreading shrub with ascending stems, growing on dry hill-sides. Leaves in whorls of 3, whitish above, prickly-pointed.

6. TAXUS. YEW.

T. bacca'ta, var. **Canadensis.** (AMERICAN YEW. GROUND HEMLOCK.) A low straggling shrub. Leaves green on both sides. *Berry globular, red.*

CLASS II. ENDOGENS OR MONOCOTYLEDONS

For characters of the Class see Part I., chap. xi.

DIVISION I. SPADICEOUS ENDOGENS.

Flowers aggregated on a *spadix* (Part I., sec. 69), with or without a *spathe* or sheathing bract.

ORDER LXXXV. ARA'CEÆ. (ARUM FAMILY.)

Herbs with pungent juice and simple or compound leaves, *these sometimes net-veined and hence* suggesting that the plants may be Exogens. Spadix usually accompanied by a spathe. Flowers either without a perianth of any kind, or with 4-6 sepals. Fruit usually a berry.

Synopsis of the Genera.

* *Leaves not linear. Flowers without perianth of any sort. Spadix accompanied by a spathe.*

1. **Ari•æ'ma.** Flowers mostly diœcious, collected on the lower part of the spadix only. Spathe (in our common species arched over the spadix. Scape from a solid bulb. Leaves compound net veined, sheathing the scape below with their petioles. Berries bright red.

2. **Ca'lla.** Flowers (at least the lower ones) *perfect*, covering the whole spadix. Spathe open and spreading, with a white upper surface, tipped with an abrupt point. Scape from a creeping rootstock. Leaves not net-veined, simple, heart-shaped.

* * *Leaves not linear. Flowers with a perianth of 4 sepals. Spadix surrounded by a spathe.*

3. **Symp'ocar'pus.** Leaves all radical, very large and veiny, appearing after the spathes which are close to the ground and are produced very early in spring. Flowers perfect, their stamens immersed in the spadix, the latter globular and surrounded by the shell-shaped spathe. Sepals hooded. Stamens 4. Fruit consisting of the soft enlarged spadix in which the seeds are sunk.

* * * *Leaves linear, sword-shaped. Spadix on the side of the scape. Flowers with a perianth of 6 sepals. No spathe.*

4. **A'corus.** Scape 2-edged, resembling the leaves, the cylindrical spadix borne on one edge. Sepals solid-wood. Stamens 6.

1. ARISÆMA. INDIAN TURNIP.

A. triphyllum. (INDIAN TURNIP.) For full description and engraving of this plant see Part I., sections 66-68.

2. CALLA. WATER ARUM.

C. palustris. (Marsh CALLA.) This plant is fully described and illustrated in Part I., section 70.

3. SYMPLOCAR'PUS. SKUNK CABBAGE.

S. fœ'tidus. Leaves 1-2 feet long, ovate or heart-shaped, short-petioled. Spathe purplish and yellowish, incurved. Plant with a skunk-like odour.—Bogs and wet places; not common northward.

4. AC'ORUS. SWEET FLAG. CALAMUS.

A. cal'amus. Scape much prolonged beyond the spadix.— Swamps and wet places.

ORDER LXXXVI. **LEMNA'CEÆ.** (DUCKWEED FAMILY.)

Very small plants floating about freely on the surface of ponds and ditches, consisting merely of a little frond with a single root or a tuft of roots from the lower surface, and producing minute monœcious flowers from a cleft in the edge of the frond. The flowers are rarely to be seen. The commonest representative with us is

Lemna polyrrhi'za, consisting of little round'sh green fronds (purplish beneath) about ¼ of an inch across, and with a *cluster* of little roots from the under surface.

ORDER LXXXVII. **TYPHA**CEÆ. (CAT-TAIL FAMILY.)

Aquatic or marsh herbs with linear sword-shaped leaves, erect or floating, and monœcious flowers, either in separate heads or on different parts of the same spike or spadix, but without a spathe and destitute of true floral envelopes. Fruit an achene, 1-seeded.

1. **Typha.** Flowers in a very dense and long cylindrical terminal spike, the upper ones staminate, the lower pistillate, the ovaries long-stalked and surrounded by copious bristles forming the down of the fruit. Leaves sword-shaped, erect, sheathing the stem below.

2. **Sparga'nium.** Flowers in separate globular heads along the upper part of the stem, the higher ones staminate, the lower pistillate, each ovary

sessile and surrounded by a few scales not unlike a calyx. Both kinds of flower staminate. Leaves flat or triangular, sheathing the stem with their bases.

1. TYPHA. CAT-TAIL FLAG.

T. latifolia. (Common CAT-TAIL.) Stem 5-8 feet high. Leaves flat. No space between the staminate and pistillate parts of the spike.—Marshy places.

2. SPARGANIUM. BUR-REED.

1. S. eurycar'pum. Stem erect, stout, 2-4 feet high. Leaves mostly flat on the upper side, keeled and hollow-sided on the lower. Heads several, panicled-spiked, the fertile an inch across in fruit. Nutlets or achenes with a broad abruptly-pointed top. —Borders of slow water and pools.

2. S. simplex, var. angustifolium. Stem slender, erect, 1-2 feet high; the leaves usually floating, long, and narrowly linear. Heads several, usually in a single row, the fertile about half an inch across. Nutlets pointed at both ends.—In slow streams.

ORDER LXXXVIII. NAIADACEÆ. (PONDWEED FAMILY.)

Immersed aquatic herbs, with jointed stems and sheathing stipules. Leaves immersed or floating. Flowers (in our common genus) perfect, in spikes or clusters, with 4 sepals, 4 stamens, and 4 ovaries; the spikes generally raised on peduncles to the top of the water. Plants of no very great interest. The most obvious characters of a few species are given here.

POTAMOGETON. PONDWEED.

1. P. natans. Submersed leaves reduced to capillary. Spikes cylindrical, above the water. Nutlets barely beaked. Floating leaves long-petioled, elliptical, with a somewhat heart-shaped base, with a blunt apex, leathery-veined.

2. P. amplifolius. Submersed leaves large, lanceolate-oval, acute at each end, on short petioles. Floating leaves large, oblong or lanceolate, or slightly cordate, long petioled, many-nerved.

3. P. lucens, var. minor. Leaves all submersed, membranous, petioled, oval or lanceolate, mucronate, shining. Stem branching.

4. **P. perfolia'tus.** Leaves all submersed, varying in width from *orbicular to lanceolate, clasping by a heart-shaped base.* Stem branching.

5. **P. pectina'tus.** Leaves all submersed, bristle-shaped. Stem repeatedly forking, filiform. *Spikes interrupted, on long slender peduncles.*

DIVISION II. PETALOIDEOUS ENDOGENS.

Flowers with a perianth coloured like a corolla.

ORDER LXXXIX. ALISMA'CEÆ. (WATER PLANTAIN F.)

Marsh herbs with flowers having 3 distinct sepals and 3 distinct petals, pistils either apocarpous or separating at maturity into distinct carpels, and hypogynous stamens 6–many. Flowers on scapes or scape-like stems. Leaves sheathing at the base, either rush-like or, when broad, mostly heart-shaped or arrow-shaped.

Synopsis of the Genera.

• *Calyx and corolla both greenish. Carpels united to the top, but separating at maturity. Leaves rush-like, fleshy.*

1. **Triglo'chin.** Flowers small, in a spike or close raceme, without bracts. Carpels when ripe splitting away from a central persistent axis.

• • *Calyx green, persistent. Corolla white. Pistil apocarpous. Leaves with distinct blades and petioles.*

2. **Alis'ma.** *Flowers perfect.* Stamens usually 6. Carpels numerous, in a ring. Leaves all radical. *Scape with whorled panicled branches.*

3. **Sagitta'ria.** *Flowers monœcious, sometimes diœcious.* Stamens numerous. Carpels numerous, in more or less globular heads. Leaves arrow-shaped, but varying greatly. Flowers mostly in whorls of 3 on the scapes, the sterile ones uppermost.

1. TRIGLOCHIN. ARROW-GRASS.

1. **T. palus'tre.** A slender rush-like plant, 6–18 inches high, found growing in bogs northward. *Carpels 3*, awl-pointed at the base, splitting away from below upwards. Spike or raceme slender, 3 or 4 inches long.

2. **T.** maritimum is also found occasionally. The whole plant is stouter than No. 1, and the carpels are usually 6 in number.

2. ALISMA. WATER PLANTAIN.

A. Planta'go. Leaves long-petioled, mostly oblong-heart-shaped, but often narrower, 3-9-nerved or ribbed, and with cross veinlets between the ribs. Flowers small, white, in a large and loose compound panicle.—Low and marshy places, often growing in the water.

3. SAGITTARIA. ARROW-HEAD.

S. varia'bilis. Very variable in size and in the shape of the leaves. Scape angled.—Common everywhere in shallow water.

ORDER XC. **HYDROCHARIDA'CEÆ.** (FROG'S-BIT F.)

Aquatic herbs, with diœcious or polygamo-diœcious flowers on scape-like peduncles from a kind of spathe of one or two leaves, the perianth in the fertile flowers of 6 pieces united below into a tube which is adherent to the ovary. Stigmas 3. Fruit ripening under water.

Synopsis of the Genera.

1. **Anach'aris.** Growing under water, the pistillate flowers alone coming to the surface. Stem leafy and branching. Perianth of the fertile flowers with a 6-lobed spreading limb, the tube prolonged to an exterior length, thread-like. Leaves crowded, pellucid, 1-nerved, sessile, whorled in threes or fours. Stamens 3-9.

2. **Vallisne'ria.** Nothing but the pistillate flowers above the surface, these on scapes of great length, and after fertilization drawn below the surface by the spiral coiling of the scapes. Tube of the perianth not prolonged. Leaves linear, thin, long and ribbon-like.

(In both genera the staminate flowers break off spontaneously and float on the surface around the pistillate ones, shedding their pollen upon them.)

1. ANACHARIS. WATER-WEED.

A. Canadensis.—Common in slow waters.

2. VALLISNERIA. TAPE-GRASS. EEL-GRASS.

V. spira'lis. **Leaves** 1-2 feet long.—Common in slow waters.

ORDER XCI. ORCHIDA'CEÆ. (ORCHIS FAMILY.)

Herbs, well marked by the peculiar arrangement of the stamens, these being gynandrous, that is, borne on or adherent to the stigma or style. There is also usually but a single stamen, of two rather widely separated anthers, but in the last genus of the following list there are 2 distinct stamens, with the rudiment of a third at the back of the stigma. As explained in Part I., Sections 71-73. the Orchids as a rule require the aid of insects to convey the *pollinia*, or pollen-masses, to the stigma, but occasionally it happens that when the anther-cells burst open the pollinia fall forward and dangle in front of the viscid stigma beneath. being sooner or later driven against it either by the wind or by the head of some insect in pursuit of honey. In all cases where the student meets with an Orchid in flower, he should, by experiment, endeavour to make himself acquainted with the method of its fertilization.

The Orchis Family is a very large one, there being probably as many as 6,000 different species. but the greater number are natives of tropical regions Many of them are *epiphytes*, or air-plants, deriving their support chiefly from the moisture of the air, through their long aerial roots which never reach the ground The perianth in many species, and particularly the *labellum*. or lip, assumes the most fantastic shapes, making the plants great favourites for hot-house cultivation. In Canada, the representatives of this great Order, though not very numerous, are among the most interesting and beautiful of our wild flowers. They are as a rule, bog-plants, and will be found in flower in early summer

Synopsis of the Genera.

* *Anther only one, but of 2 cells*,

← *Lip with a spur*

1. **Orchis.** The 3 sepals and 2 of the petals form a hood over the centre of the flower ; the lip turned down. Anther a single gland at the base of the pollen stigma. Leaves 2,

2. Habenaria.

← ← *Lip without a spur*.

3. **Spiranthes.** twisted. long. the *and with* each side base.

4. Goodyera. Flowers very much as in Spiranthes, *but the lip sac-shaped,*
and without a prolongation at the base. Leaves white veiny, in a tuft at
the base of the scape.

♦ ♦ ♦ *Lip without a spur. Anther on the apex of the style, its cell like a lid.*
Pollen-masses 4, soft, separate. without an attached stalk or gland.

5. Calopogon. Ovary not twisted. the lip consequently turned *towards the*
stem. Flowers large, pink-purple, 2–6 at the top of the scape ; the
lip spreading at the outer end and a beautifully bearded above with coloured hairs. Leaf grass-like, only one.

6. **Calyp'so.** Flower solitary, large, showy, variegated with purple, pink,
and yellow. Lip large, inflate , sac-shaped, 2 pointed under the apex.
Scape short, from a solid bulb, with a single ovate or slightly heart-
shaped leaf below.

7. Corallorhiza. Root a branched collection with the small dull flowers
in spikes on scape, without green leaves, or have mere sheaths instead of
leaves. Roots thick branched and coral like. *Perianth gibbous or*
slightly spurred at base. Lip with 2 ridges on the inner part of the face.

♦ ♦ *Anthers 2, one on each side of the stigma, and a triangular body, which is*
the rudiment of a third, at the back of the stigma. Pollen loose and
powdery or pulpy.

8. Cypripedium. Lip a large inflated sac, into the mouth of which the
style is bended. 4 petals and the other petals much like the anthers
appears, only 2, united their filaments generally united into one under
the lip. Leaves large, many nerved. Flowers solitary or few.

1. ORCHIS, ORCHIS.

O. specta'bilis. (SNOWY ORCHIS.) Scape 4-angled, 4–7 inches
high, bearing a few flowers in a spike. The arching upper lip
pink-purple, the *labellum* white ; each flower in the axil of a leaf-
like bract.—Rich woods.

2. HABENARIA. REIN ORCHIS.

1. H. tridentata. *Scale few-flowered,* the flowers very small
greenish-white. *Lip wedge-shaped, truncate and 3-toothed at the*
apex. Spur slender, longer than the ovary, curved upwards.
Stem less than a foot high, etc., etc., with one oblanceolate leaf
below and 2 or 3 or 4 smaller bract-leaves.—Wet woods.

2. H. viridis, var. bracteata. *Spike many-flowered,* the
flowers whitish-green. *Lip oblong, obtuse, 2-3 lobed at the tip,*
much longer than the very short and sac-like spur. Stem 6–12
inches high, leafy, the lower leaves obovate, the upper oblong or
lanceolate, gradually reduced to bracts much longer than the
flowers.

3. **H. hyperbo'rea.** Spike many-flowered, *long and dense.* Flowers small, greenish. Lip lanceolate, *entire, about the same length as the slender incurved spur.* Stem 6–24 inches high, *very leafy, the leaves lanceolate and erect,* and the bracts longer than the flowers.—Bogs and wet woods.

4. **H. dilata'ta.** Not unlike No. 3, but more slender and with *linear leaves* and *white flowers.*

5. **H. rotundifo'lia.** Spike few-flowered, loose. Flowers rose-purple, *the lip usually white, spotted with purple,* 3-lobed, the middle lobe larger and notched, longer than the slender spur. Stem 5–9 inches high, *naked and scape-like above, bearing a single roundish leaf* at the base.—Bogs and wet woods.

6. **H. Hook'eri.** Spike many-flowered, *strict.* Flowers yellowish-green, the lip lanceolate, pointed, incurved ; petals lance-awl-shaped. *Spur slender, acute, nearly an inch long.* Stem scape-like above, *2-leaved at the base, the leaves orbicular.*—Woods.

7. **H. orbicula'ta.** Spike many-flowered, *loose and spreading.* Flowers *greenish-white.* Lip narrowly linear, obtuse. *Spur curved, more than an inch long, thickened towards the apex.* Scape 2-leaved at the base, the leaves *very large, orbicular, and lying flat on the ground,* shining above, silvery beneath.—Rich woods.

8. **H. blephariglot'tis.** (WHITE FRINGED ORCHIS.) Spike many-flowered, open. *Flowers white,* very handsome ; *the lip fringed, but not lobed,* at the apex. *Spur thread-shaped, three times as long as the lip.* Stem a foot high, leafy ; the leaves oblong or lanceolate, the bracts shorter than the ovaries.—Peat-bogs, &c.

9. **H. leucophæa.** (GREENISH FRINGED ORCHIS.) Spike as in the last, but the flowers *greenish* or *yellowish-white.* Lip 3-parted above the stalk-like base, the *division-fringed.* Spur gradually thickened downward, longer than the ovary. Stem leafy, 2–4 feet high. Leaves oblong-lanceolate ; bracts a little shorter than the flowers.—Wet meadows.

10. **H. psyco'des.** (PURPLE FRINGED ORCHIS.) Spike cylindrical, many-flowered, the *flowers pink-purple,* fragrant. Lip fan-shaped, 3-parted above the stalk-like base, *the divisions fringed.* Spur curved, somewhat thickened downward, very long.

3. SPIRANTHES. LADIES' TRESSES.

1. **S. Romanzovia'na.** Spike dense, oblong or cylindrical. Flowers *pure white, in 3 ranks in the spike.* Lip ovate-oblong, contracted below the wavy recurved apex. Stem 5-15 inches high, leafy below, leafy-bracted above; the leaves oblong-lanceolate or linear.—Cool bogs.

2. **S. gra'cilis.** *Flowers in a single spirally twisted rank* at the summit of the very slender scape. Leaves with blades all in a cluster at the base, ovate or oblong. Scape 8-18 inches high.— Sandy plains and pine barrens.

4. GOODYERA. RATTLESNAKE-PLANTAIN.

1. **G. repens.** *Flowers in a loose 1-sided spike.* Lip with a recurved tip. Scape 5-8 inches high, Leaves thickish, petioled, intersected with whitish veins.—Woods, usually under evergreens.

2. **G. pubes'cens.** *Spike not 1-sided.* Plant rather larger than the last, and the leaves more strongly white-veined.—Rich woods.

5. CALOPOGON. CALOPOGON.

C. pulchel'lus. Leaf linear. Scape a foot high. Flowers an inch across.—Bogs.

6. CALYPSO. CALYPSO.

C. borea'lis. A beautiful little plant growing in mossy bogs. The lip woolly inside; the petals and sepals resembling each other, lanceolate, sharp-pointed. Column winged.

7. CORALLORHIZA. CORAL-ROOT.

1. **C. inna'ta.** Flowers small; the lip whitish or purplish, often crimson-spotted, 3-lobed above the base. Spur very small. Stem slender, brownish-yellow, with a few-flowered spike.— Swamps.

2. C. multiflo'ra. Spike many-flowered. Stem purplish, stout. Lip deeply 3-lobed. Spur more prominent than in No. 1.—Dry woods.

3. C. Macrae'i. Spike crowded, of numerous large flowers, all the parts of the perianth streaked and marked with dark lines. Lip not lobed. Spur none, but the base of the perianth gibbous.

—This species is probably not common. It has, however, been found at Barrie, Ont., and is reported also from the south-western part of the province.

8. CYPRIPE'DIUM. LADY'S SLIPPER. MOCCASON-FLOWER.

1. **C. parviflo'rum.** (SMALLER YELLOW LADY'S SLIPPER.) Stem leafy to the top, 1–3 flowered. Lip yellow, *flattish above*, rather less than an inch long. Sepals and petals wavy-twisted, brownish, pointed, longer than the lip.—Bogs and wet woods.

2. **C. pubes'cens.** (LARGER YELLOW L.) Lip flattened *laterally*, rounded above, larger than in No. 1, but the two species are not sufficiently distinct.

3. **C. specta'bile.** (SHOWY L.) *Lip very large, white, pinkish in front.* Sepals and petals *rounded, white*, not longer than the lip.—Bogs.

4. **C. acau'le.** (STEMLESS L.) *Scape naked*, 2-leaved *at the base, 1-flowered.* *Lip rose-purple*, split down the whole length in front, veiny. Sepals and petals greenish.—Dry or moist woods, under evergreens.

ORDER XCII. **IRIDA'CEÆ.** (IRIS FAMILY.)

Herbs with equitant leaves and perfect flowers. The 6 petal-like divisions of the perianth in 2 (similar or dissimilar) sets of 3 each; the tube adherent to the 3-celled ovary. Stamens 3, distinct or monadelphous, opposite the 3 stigmas, and with anthers **extrorse**, that is, on the outside of the filaments, facing the divisions of the perianth and opening on that side. Flowers from leafy bracts.

Synopsis of the Genera.

1. **Iris.** The 3 outer divisions of the perianth recurved, the 3 inner erect and smaller. Stamens distinct, the anthers sessile and under a flat and petal-like division of the style. The tube adherent to the tube of the perianth. Flowers large. Leaves sword-shaped or linear.

2. **Sisyrin'chium.** The 6 divisions of the perianth alike, spreading. Stamens monadelphous. Stigmas three filiform. Flowers small. Stems 2-edged. Leaves grass-like. Flowers blue, clustered, from a leafy bract. Flat low and slender.

1. IRIS. FLOWER-DE-LUCE.

1. I. versic'olor. (LARGER BLUE FLAG.) Stem stout and leafy, from a thickened rootstock. Leaves sword-shaped. Flowers violet-blue, 2 or 3 inches long. Inner petals much smaller than the outer. — Wet places.

2. SISYRINCHIUM. BLUE-EYED GRASS.

S. Bermudia'na. A pretty little plant, rather common in moist meadows among grass. The divisions of the delicate blue perianth obovate, notched at the end, and bristle-pointed from the notch. Roots fibrous.

ORDER XCIII. DIOSCOREA'CEÆ. (YAM FAMILY.)

Represented with us by the genus

DIOSCOREA. YAM.

D. villo'sa. (WILD YAM-ROOT.) A slender twiner with knotted rootstocks, and net-veined, heart-shaped, 9–11-ribbed, petioled leaves. Flowers diœcious, small, in axillary racemes. Stamens 6. Pod with three large wings. — Reported only from the warm and sheltered valley lying between Hamilton and Dundas, Ont.

ORDER XCIV. SMILA'CEÆ. (SMILAX FAMILY.)

Climbing plants, more or less shrubby, with alternate *ribbed and net-veined* petioled leaves, and small diœcious flowers in umbels. Perianth regular, of 6 greenish sepals, free from the ovary. Stamens as many as the sepals, with 1-celled anthers. Ovary 3-celled, surmounted by 3 sessile spreading stigmas. Fruit a small berry. Represented by the single genus

SMILAX. GREENBRIER. CAT-BRIER.

1. S. his'pida. *Stem below densely covered with long weak prickles.* Leaves large, ovate or heart-shaped, pointed, thin, 5–9 nerved. Peduncles of the axillary umbels much longer than the petioles. Berry black. — Moist thickets.

2. S. herba'cea. (CARRION-FLOWER.) Stem herbaceous, *not prickly.* Leaves ovate-oblong and heart-shaped, 7–9-ribbed, long-petioled, mucronate. Flowers carrion-scented. Berry bluish-black. — Meadows and river-banks.

ORDER XCV. **LILIA'CEÆ.** (LILY FAMILY.)

Herbs, distinguished as a whole by their regular and symmetrical flowers, having a 6-leaved perianth (but 4-leaved in one species of Smilacina) free from the usually 3-celled ovary, and as many stamens as divisions of the perianth (*one before each*) with 2-celled anthers. Fruit a pod or berry, generally 3-celled. The outer and inner divisions of the perianth coloured alike, except in the genus Trillium. (See part I., sections 61–65, for description of typical plants of this Order.)

Synopsis of the Genera.

* *Leaves net-veined, all in one or two whorls. The stem otherwise naked, rising from a fleshy rootstock. Styles 3.*

1. **Trillium** Leaves 3 in a whorl at the top of the stem. Divisions of the perianth in 2 sets, the outer green, the inner coloured. (See Part I, sections 64 and 65.)

2. **Medeola.** Leaves in 2 whorls, the lower near the middle of the stem, and consisting of 5-9 leaves, the upper of (generally) 3 small leaves, near the summit. Stem tall, covered with loose wool. Flowers small, in an umbel. Divisions of the perianth alike, greenish-yellow, recurved. Anthers turned outwards. Styles thread-shaped. Berry globular or nearly so, dark purple.

* * *Leaves straight-veined, linear, grass-like, alternate. Stem simple, rising (in our species) from a coated bulb. Styles 3.*

3. **Zygade'nus.** Flowers perfect or polygamous, greenish-white, in a few-flowered panicle ; the divisions of the perianth each with a conspicuous obcordate spot or gland on the inside, near the narrowing base. Stem smooth and glaucous.

* * * *Leaves straight-veined, but   alternate. Stem from a rootstock or fibrous   from a bulb. Style   at the base, but more   .*

     ← *Perianth of   pieces (polyphyllous).*

4. **Uvula'ria.** Stem leafy, forking above. Flowers yellow, at least an inch long, drooping, lily-like, usually solitary (but occasionally in pairs) at the end or in the forks of the stem. Style deeply 3-cleft. Pod triangular. Leaves clasping-perfoliate.

5. **Clinto'nia.** Stemless, the naked scape sheathed at the base by 2, 3, or 4 large oblong or oval ciliate leaves. Flowers few, drooping, in an umbel at the top of the scape. Filaments long and slender. Style long, the stigmas hardly separate. Berry blue.

6 . . . reptopus. Stem leafy and forking. Flowers small, not quite in the
axils of the ovate clasping leaves, *on slender peduncles which are abruptly bent near the middle.* Anthers arrow-shaped, 2-horned *at th apex.*

← ← *Perianth of one piece (monophyllous)*

7. Smilacina. Flowers small, white, in a terminal raceme. Perianth 6 parted . . . . parted in one species, spreading. Style short and thick Stigma obscurely lobed. Filaments slender.

8. Po ygonatum . Flowers small, greenish, nodding, *mostly in pairs* in the axils of the nearly sessile leaves. Perianth cylindrical, 6-lobed at the summit, *the 6 stamens inserted on the tube above the middle.* Stem simple, from a long and knotted rootstock. Leaves glaucous beneath.

. . . . *Leaves* . . . . *not* . . . . Stem from a coated or . . . *bulb* . . . . the stigma sometimes . . . . *Fruit a . . . . the partitions (loculicidal).*

9. Lilium. . . . a . . . . bulb, the leaves often whorled or crowded. Anthers at first erect, at last versatile. Style long, rather club-shaped. Stigma 3-lobed. Pod . . . . Flowers large and showy, one or more.

10. Erythronium. For full description see Part I. sections 61-63. (Dog's-tooth Violet.)

11. Allium. Scape naked, from a coated bulb. The radical leaves broad and flat, withering bef re the flowers are developed. Flowers white, in an umbel. Style thread-like. Strong-scented plants.

1. TRILLIUM. WAKE R BIN

1. T. grandiflo'rum. (LARGE WHITE TRILLIUM.) *Leaves sessile, longer than broad.* Petals white (rose-coloured when old), obovate.—Rich woods.

2. T. erectum. (PURPLE TRILLIUM.) *Leaves sessile, about as broad as long.* Petals dull purple, *ovate.*—Rich woods. Var. album, with greenish-white petals, is found along with the purple form. It does not appear to be clearly distinguished from No. 1

3. T. erythrocarpum (PAINTED TRILLIUM.) *Leaves distinctly petioled, rounded at the base.* Petals pointed, white, with purple stripes inside at the base. Not uncommon northward in damp woods and low grounds.

2. MEDIOLA. INDIAN CUCUMBER-ROOT.

M. Virgin'ica. Stem 1-3 feet high.—Rich woods.

3. ZYGADE'NUS. ZYGADENE.

Z. glaucus. Not uncommon in bogs and beaver-meadows northward. Leaves flat and pale.

4. UVULA'RIA. BELLWORT.

U. grandiflo'ra.—Rich woods.

5. CLINTONIA. CLINTONIA.

C. borea'lis. Umbel 2-7-flowered. Leaves 5-8 inches long. Perianth pubescent outside.—Damp woods, often under evergreens.

6. STREP'TOPUS. TWISTED-STALK.

S. ro'seus. Flowers rose-purple.—Damp woods.

7. SMILACINA. FALSE SOLOMON'S SEAL.

1. **S. racemo'sa.** (FALSE SPIKENARD.) *Raceme compound.* Stem pubescent, 2 feet high. Leaves many, oblong, taper-pointed, ciliate. Berries speckled with purple.—Rich woods and thickets.

2. **S. stella'ta.** *Raceme simple.* Stem nearly smooth, 1-2 feet high. Leaves 7-12, oblong-lanceolate, slightly clasping. Berries black.—Moist woods and copses.

3. **S. trifo'lia.** *Raceme simple.* Stem low (3-6 inches), glabrous. Leaves usually 3, oblong, the bases sheathing. Berries red.—Bogs.

4. **S. bifo'lia.** Distinguished at once by the *4-parted perianth and the 4 stamens.* Raceme simple. Stem 3-5 inches high. Leaves usually 2, but sometimes 3.—Moist woods.

8. POLYGONATUM. SOLOMON'S SEAL.

P. biflo'rum. (SMALLER SOLOMON'S SEAL.) Stem slender, 1-3 feet high. Leaves ovate-oblong or lance-oblong. Filaments hairy.—Rich woods.

9. LILIUM. LILY.

1. **L. Philadel'phicum.** (WILD ORANGE-RED LILY.) Divisions of the perianth *narrowed into claws* below, not recurved at the top. Flowers *erect*, 1-3, orange, spotted with purple inside. Leaves linear-lanceolate, the upper mostly in whorls of 5-8.— Sandy soil.

2 **L. Canaden'se.** (WILD YELLOW LILY.) Divisions of the perianth *recurved above the middle.* Flowers *nodding*, few, orange, spotted with brown inside. Leaves remotely whorled, 3-ribbed.—Swamps and wet meadows.

3. **L. super'bum.** (TURK'S-CAP LILY.) Divisions of the perianth *very strongly recurved.* Flowers nodding, *often numerous,* in a pyramidal raceme, bright orange, dark-purple-spotted within. Lower leaves whorled, 3-ribbed or nerved. Stem taller than either of the first two, 3-7 feet.—Rich low grounds, commoner southward and south-westward.

10. ERYTHRO'NIUM. Dog's-tooth Violet.

E. America'num. (YELLOW ADDER'S TONGUE.) Perianth light yellow, sometimes spotted at the base.—Copses and rich meadows.

11. ALLIUM. ONION. LEEK.

A. tricoccum. (WILD LEEK.) Leaves lance-oblong, 5-9 inches long, 1-2 inches wide. Pod strongly 3-lobed. Scape 9 inches high.—Rich woods.

ORDER XCVI. JUNCA'CEÆ. (RUSH FAMILY.)

Grass-like or sedge-like plants, with, however, flowers similar in structure to those of the last Order. Perianth greenish and glumaceous, of 6 divisions in 2 sets of 3 each. Stamens 6 (occasionally 3). Style 1. Stigmas 3. Pod 3-celled, or 1-celled with 3 placentæ on the walls. The plants of the Order are not of any very great interest to the young student, and the determination of the species is rather difficult. A brief description of a few of the most common is given here, as an easy introduction to the study of the Order with the aid of more advanced text-books.

Synopsis of the Genera.

1. **Lu'zula.** Plant less than a foot high. Leaves linear or lance-linear, flat, *usually hairy. Pod 1-celled, 3-seeded.* Flowers in umbels or in spikes. Plants usually growing in *dry* ground.

2. **Juncus.** Plants *always smooth,* growing in water or *wet* soil. Flowers small, greenish or brownish, panicled or clustered. Pod 3-celled, *many-seeded.*

1. LU'ZU_A. WOOD-RUSH.

1. **L. pilo'sa.** Flowers umbelled, long-peduncled, brown-coloured. Sepals pointed.—Shady banks.

2. **L. campestris** has the flowers (light brown) in 4–12 spikes, the spikes umbelled. Sepals bristle-pointed.—Fields and woods.

2. JUNCUS. RUSH.

1. **J. effu'sus.** (COMMON OR SOFT RUSH.) Scape 2–4 feet high, soft and pliant, furnished at the base with merely *leafless sheaths*, the inner sheaths awned. The many-flowered panicle sessile, apparently produced from the *side* of the scape, owing to the involucral leaf being similar to and continuing the scape. Flowers small, greenish, *only 1 on each pedicel.* Stamens *3.* Pod triangular-obovate, *not pointed.*—Marshes.

2. **J. filifor'mis** has a very slender scape (1–2 feet high), fewer flowers than No. 1, and *6 stamens* in each. Pod broadly ovate and *short-pointed. No leaves.*

3. **J. bufo'nius.** Stem *leafy*, slender, 3–9 inches high, branching from the base. Panicle terminal, spreading. Flowers single on the *pedicels.* Sepals awl-pointed, the outer set much longer than the inner. *Stamens 6.*—Ditches along road-sides.

4. **J. ten'uis.** Stems *leafy below*, wiry, 9–18 inches high, *simple*, tufted. Panicle loose, shorter than the slender involucral leaves. Flowers greenish, single on the pedicels ; the sepals longer than the blunt pod.—Open low grounds.

ORDER XCVII. **PONTEDERIA'CEÆ.** (PICKEREL-WEED FAMILY.)

The most common representative of this Order with us is

PÒNTEDE'RIA. PICKEREL-WEED.

P. corda'ta. A stout plant growing in shallow water, sending up a scape bearing a single large arrow-heart-shaped blunt leaf, and *a spike of violet-blue flowers with a spathe-like bract.* Perianth 2-lipped, the 3 upper divisions united, the 3 lower spreading, the whole revolute-coiled after flowering, **the** fleshy base enclosing the fruit. Stamens 6, 3 of them exserted on long filaments, the rest short.

ORDER XCVIII. **ERIOCAULONA'CEÆ.** (PIPEWORT FAMILY.)

Represented with us by the genus

ERIOCAU'LON. PIPEWORT.

E. septangula're. A slender plant with a naked scape 2-6 inches high, growing in shallow water in the margins of our northern ponds. Leaves short, awl-shaped, in a tuft at the base. Flowers in a small woolly head at the summit of the scape, monœcious. Perianth double ; the outer set or calyx of 2-3 keeled sepals ; the corolla tubular in the sterile flowers and of 2-3 separate petals in the fertile ones. Scape 7-angled. The head (except the beard) lead-coloured.

DIVISION III. GLUMACEOUS ENDOGENS.

Flowers without a proper perianth, but subtended by thin scales called *glumes.*

This Division includes two very large Orders—Cyperaceæ and Gramineæ—both of which present many difficulties to the beginner. Accordingly no attempt will be made here to enumerate and describe all the commonly occurring species of these Orders. It will be sufficient for the purposes of this work to describe two or three of the very commonest representatives of each, so as to put the beginner in a position to continue his study of them with the aid of Gray's Manual or other advanced work.

ORDER CXIX. **CYPERA**CEÆ. (SEDGE FAMILY.)

Grass-like or rush-like herbs, easily distinguished from Grasses **by the** sheaths of the leaves, which in the Sedges are closed round the culm, not split. Flowers in spikes, each flower in the axil of a glume-like bract, either altogether without a perianth **or** with a few bristles or **scales** inserted below the ovary. Ovary 1-celled, becoming **an** achene (2- or 3-angled). Style **2- or** 3-cleft. Stamens mostly 3, occasionally 2.

We shall describe one species of each of five genera.

1. Cyperus diandrus.

This plant (Fig. 1) is from 4 to 10 inches in height.   The culm is *triangular*, leafy towards the base, but naked above.   At the summit there is an umbel the rays of which are unequal in length, and on each ray are clustered several *flat brown-colour-ed* spikes, the scales of which are imbricated in two distinct rows. At the base of the umbel there are 3 leaves of very un-equal length, forming a sort of involucre, and the base of each ray of the umbel is sheathed.   In each spike every scale ex-cept the lowest one con-tains a flower in its axil. The flower (Figs. 2 and 3) is entirely destitute of perianth, and consists of 2 *stamens and an ovary sur-mounted by a 2-cleft style,* being consequently *perfect.* —The plant is pretty easily met with in low wet places.

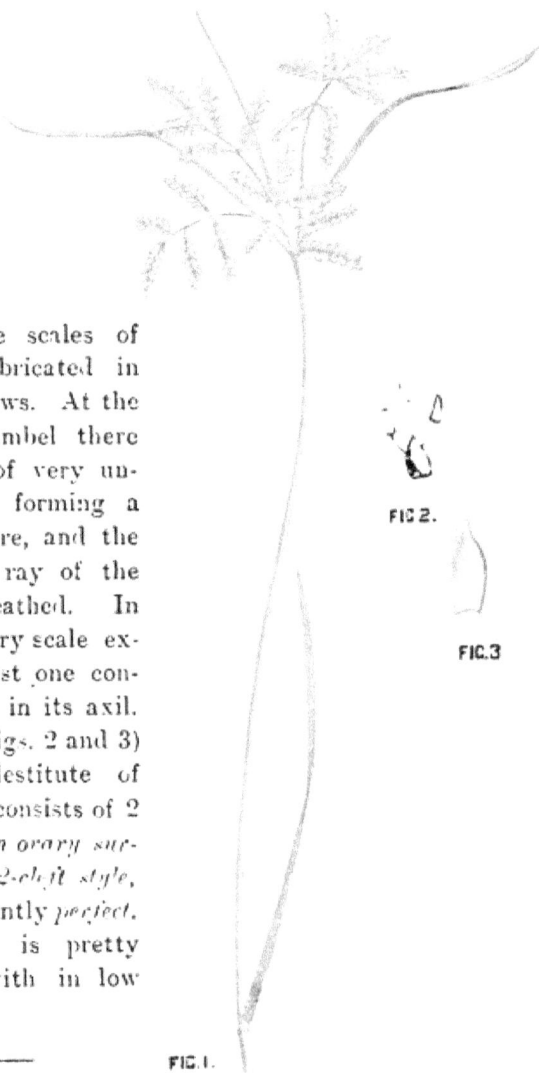

FIG. 2.

FIG. 3.

FIG. 1.

2. Eleoch'aris obtu'sa.

In this plant, which grows in muddy soil in tufts 8 to 14 inches in height, there is but a single spike at the summit of each slender culm, and the scales of the spikes, instead of being imbricated in 2 rows and thus producing a flat form, are *imbricated all round*. The scales are very thin in texture, with a midrib somewhat thicker, and are usually brownish in colour. Each of them contains a perfect flower in its axil. Instead of a perianth there are *6 or 8 hypogynous barbed bristles*. The stamens (as is generally the case in this Order) are 3 in number, and the style is usually 3-cleft. Observe that the style is enlarged into a sort of bulb at the base, *this bulbous portion persisting as a flattish tubercle on the apex of the achene*. The culms are without leaves, being merely sheathed at the base.

3. Scirpus pungens.

A stout marsh-plant, 2 or 3 feet high, with a sharply triangular hollow-sided culm, and bearing at the base from 1 to 3 channelled or boat-shaped leaves. The rusty-looking spikes vary in number from 1 to 6, and are in a single sessile cluster which appears to spring from the side of the culm, owing to the 1-leaved involucre resembling the culm and seeming to be a prolongation of it. Each scale of the spike is 2-cleft at the apex, and bears a point in the cleft. The flowers are perfect, with 2 to 6 bristles instead of perianth, 3 stamens, and a 2-cleft style, *but there is no tubercle on the apex of the achene*. The culms of this plant spring from stout running rootstocks.

4. Eriophorum polystach'yon.

A common bog-plant in the northern parts of Canada, resembling Scirpus in the details as to spikes, scales, &c., but differing chiefly in this, that the bristles of the flowers are very delicate and become very long after flowering, so that the spike *in fruit looks like a tuft of cotton*. The culm of our plant is triangular, though not manifestly so, and its leaves are hardly, if at all, channelled. The spikes are several in number, and are on nodding peduncles, and the involucre consists of 2 or 3 leaves. Culm 15 or 20 inches high.

5. Carex intumes'cens.

The species of the genus Carex are exceedingly numerous and difficult of study. The one we have selected (Fig. 4) is one of the commonest and at the same time one of the easiest to examine. In this genus the flowers are monœcious, the separate kinds being either borne in different parts of the same spike, or in different spikes. The genus is distinguished from all the others of this Order by the fact of the achene *being enclosed in a bottle-shaped more or less inflated sac*, which is made by the union of the edges of two inner bractlets or scales. To this peculiar sac (Figs. 5 and 6) which encloses the achene the name *perigynium* is given. The culms are always triangular and the leaves grass-like, usually roughened on the margins and on the keel.

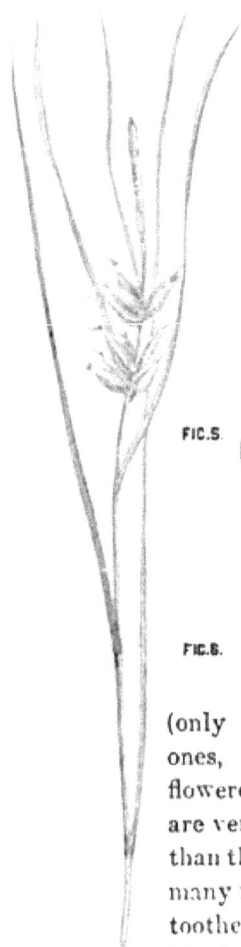

In the species under examination (which may be found in almost any wet meadow) the culm is some 18 inches high. The staminate spike (only one) is separate from and *above* the fertile ones, which are 2 or 3 in number, few-(5 to 8) flowered, and quite near together. The perigynia are very much inflated, that is, very much larger than the achenes ; they are distinctly marked with many nerves, and taper gradually into a long 2-toothed beak from which protrude the 3 stigmas. The bracts which subtend the spikes are leaf-like, and extend much beyond the top of the culm.

FIG.5

FIG.6

FIG.4

ORDER CXX. GRAMIN'EÆ. (GRASS FAMILY.)

Herbs somewhat resembling those of the last Order, but the culms are hollow except at the joints, and the sheaths of the leaves are split on the opposite side of the culm from the blade. The student is referred to Part I, section 74, for the description and illustration of a Grass-flower. In addition to the terms there defined it may be explained that the name *ligule* is given to a thin membranaceous upward extension of the sheath, and *lodicules* to some minute hypogynous scales usually accompanying each flower.

We shall give brief descriptions of representatives of six common Canadian genera.

1. Agros'tis vulga'ris. (RED-TOP.)

In the examination of Timothy it was found that the very numerous flowers were so densely crowded together as to form a cylindrical spike. In the well-known Grass now under consideration the flowers form a loose open panicle. As in Timothy, *each pair of glumes encloses but one flower*, and we must observe that the term *spikelet*, so far as Grasses are concerned, is applied to the pair of glumes and whatever is contained in them, whether one flower, or many, as is often the case. In Red-top and Timothy *the spikelets are 1-flowered*. The culm of our Grass is from 1 to 2 feet high, and the whole panicle has a purple appearance. Observe the very thin texture of the *palets*, and also that one of them (the lower, *i.e.*, the one farthest from the stalk) is nearly twice as large as the other, and is marked with 3 nerves.

2. Poa pratensis. (COMMON MEADOW-GRASS.)

The inflorescence of this very common Grass (Fig. 7) is a greenish panicle. The spikelets (Fig. 8) contain from 3 to 5 flowers and are laterally compressed. The *glumes* are the lowest pair of scales, and they are generally shorter than the flowers within them. Observe the delicate whitish margin of the lower palet of each flower (Fig. 9), and the thin texture of the upper one. Count also, if you can, the five nerves on the lower palet, and observe the 2 teeth at the apex of the upper one.

3. Bromus secalinus. (Chess.)

A common pest in wheat fields. This Grass is comparatively easy of examination on account of the size of the spikelets and flowers. The spikelets form a spreading panicle, each of them being on a long slender nodding pedicel, and containing from 8 to 10 flowers. Of the 2 glumes at the base of each spikelet one is considerably larger than the other. The outer or lower palet of each flower is tipped with a bristle, while *the upper pa'et at length becomes a'tached to the groove of the oblong grain.* Observe that the *glumes* are not awned.

4. Trit'icum repens. (COUCH-GRASS.)

Very common in cultivated grounds. In this Grass the spikelets are sessile on opposite sides of the zigzag peduncle, so that the whole forms a spike. Each spikelet is 4 to 8-flowered, and there is but one at each joint of the peduncle, the *side* of the spikelet being against the stalk. The glumes are nearly equal in size, and the lower palet of each flower closely resembles the glumes, but is sharp-pointed or awned. The Grass spreads rapidly by running root-stocks, and is troublesome to eradicate.

5. Pan icum capilla're.

(OLD-WITCH GRASS.)

This Grass is to be found everywhere in sandy soil and in cultivated grounds. The sheaths and the leaves are very hairy, and the panicle very large, compound, and loose, the pedicels being extremely slender. The culm is from 10 to 15 inches high. Of the 2 glumes one is much larger than the other. Unless you are careful you will regard the spikelets as 1-flowered ; observe, however, that in addition to

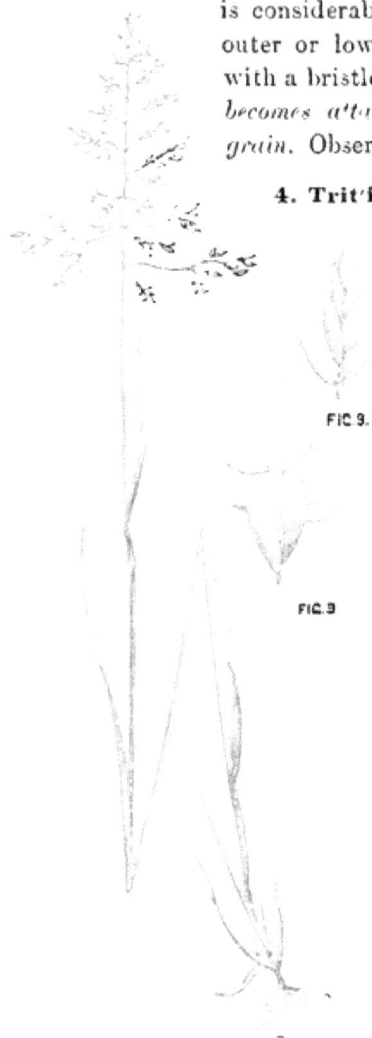

FIC 9.

FIC. 9

FIC. 7.

the one manifestly perfect flower *there is an extra palet below*. This palet (which is very much like the larger glume) is a rudimentary or abortive second flower, and the spikelet may be described as 1½-flowered.

6. Pan'icum Crus-galli. (Barnyard-Grass.)

This is a stout coarse Grass, common in manured soil. The culms are from 1 to 4 feet in height, and branch from the base. The spikelets form dense spikes, and these are crowded in a dense panicle which is rough with stiff hairs. The structure of the spikelets is much the same as in No. 5, but the palet of the neutral flower is pointed with a rough awn.

7. Seta'ria glau'ca. (Foxtail.)

Here the inflorescence is apparently a dense bristly cylindrical spike. In reality, however, it is a spiked panicle, the spikelets being much the same as in Panicum, but their *pedicels* are prolonged beyond them into awn-like bristles. In this plant the bristles are in clusters and are barbed upwards. *The spikes are tawny-yellow in colour.*

SERIES II.

FLOWERLESS OR CRYPTOGAMOUS PLANTS.

PLANTS not producing true flowers, but reproducing themselves by means of spores instead of seeds, the spores consisting merely of simple cells, and not containing an embryo.

In the introductory part of this work no reference was made to the plants of this series, chiefly because the examination of them is attended with too much difficulty for the young beginner. It is true that the structure of the Cryptogams is less complicated than that of flowering plants, but the organs requiring examination are so minute as to put a proper understanding of their nature beyond the reach of any but practised observers. Besides, there are many details of structure and function with which botanists are as yet but imperfectly acquainted, so that on the whole the better plan is that which has been adopted, viz.: to study first those forms which are better understood, and which do not require so great a nicety of observation, and then to make an effort to understand the relation between these lower forms which are now to occupy our attention, and those with which our previous practice has made us more or less familiar.

146

The series of Cryptogamous plants is subdivided into three classes, as follows :—

1. ACROGENS.
2. ANOPHYTES.
3. THALLOPHYTES.

The Acrogens, the only one of these three classes to which we shall devote any special attention here, derive their name from the mode of growth of the stem, which is quite different from that of Exogens and Endogens. In the Acrogens, or *point-growers*, the stem is increased by *successive additions to its extremity* only, all the tissues below this being completed when they are first formed, and undergoing no subsequent change. This class embraces the Ferns, Horsetails, and Club-mosses, plants which in addition to the peculiar mode of growth of the stem are characterized among Cryptogams by the presence of *vascular* as well as cellular tissue in their composition. The plants of the other two classes, the Anophytes and the Thallophytes, are composed of cellular tissue only.

The Anophytes include the true Mosses and Liverworts, which are like the Acrogens in their mode of growth, but, as just stated, are without any woody tissue whatever.

The Thallophytes include the lowest plants of all, such as Sea-Weeds, Lichens, Mushrooms and Moulds (Fungi). All these plants fail to exhibit any distinction of stem and leaf; they consist merely of an irregular mass of cellular tissue, the simplest ones of all being reduced to a single cell, a state of things well exemplified in the microscopic plant known as *Red Snow* (these would and Bread-mould consist of a number of cells placed end to end).

In all the Cryptogams reproduction is carried on by means of *spores*. These are extremely minute bodies, somewhat similar in structure to a pollen-grain, being provided with a double coat. In germination the inner coat is protruded and forms a thin green leaf-like expansion, with very minute root-fibres on the lower side. On the same side are also produced little cellular bodies of two distinct sorts, corresponding to the stamens and carpels of Phanerogams, and fertilization takes place in a manner analogous to that observed in the action of pollen. As a result of fertilization a new plant is produced resembling that which produced the spore. It is to be noticed, therefore, that while a true seed contains in itself the embryo of the new plant, which is directly produced in the process of germination, a spore produces a body on which are *afterwards* developed the organs the mutual action of which gives rise to the new plant.

As the spores do not contain an embryo, there is of course nothing answering to the cotyledons and the radicle with which we are familiar in Phanerogamous plants. Cryptogams are therefore also known as *acotyledonous* plants, or shortly *acotyledons*.

FERNS.

These beautiful plants are favourites everywhere, and we shall therefore enter into a description of their characteristics with sufficient minuteness to enable the young student to determine with tolerable certainty the names of such representatives of the Family as he is likely to meet with commonly.

Fig. 10 is a representation of the common Polypod Fern. It may be found in shady places almost everywhere, growing for the most part on rocks. The horizontal stem, shown in the

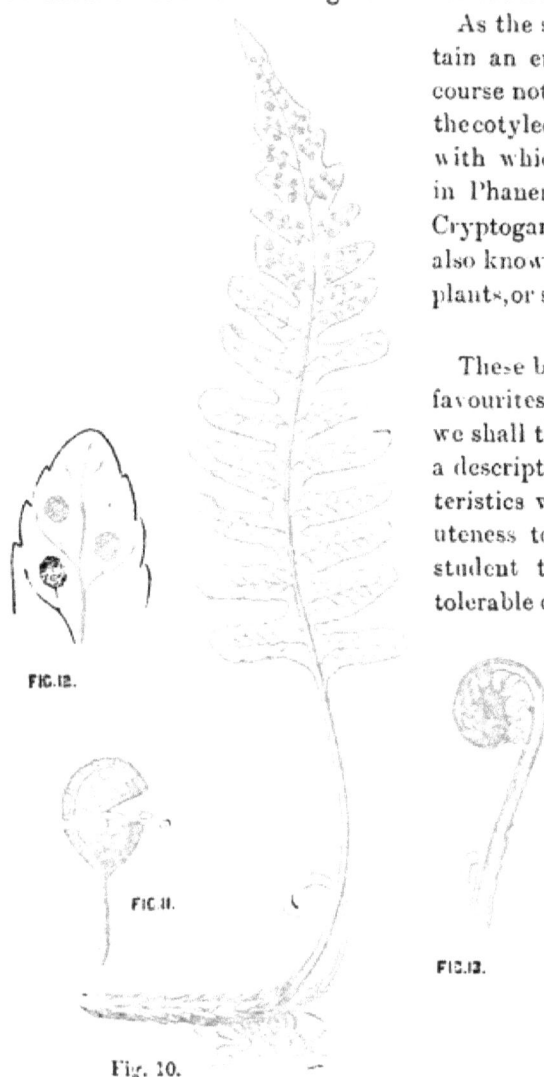

FIG.12.

FIC.II.

FIG.13.

Fig. 10.

lower part of the figure, is a *rhizome*, which runs along beneath the surface of the ground, the fibrous roots being produced from the lower side.   From the upper side of the rhizome grows the upright leaf, with a long petiole and pinnately-lobed blade. It is, however, something more than an ordinary leaf.   On the back of the upper lobes
(the figure shows the back) you observe rows
of dots on each side of the middle vein.
These dots are clusters of *spore-cases* or
sporangia. The *clus-
ters* are called *sori*.
The microscope
shows each sporan-
gium or spore-case
to be an almost

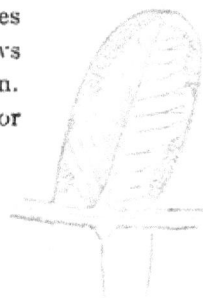

FIC.15.

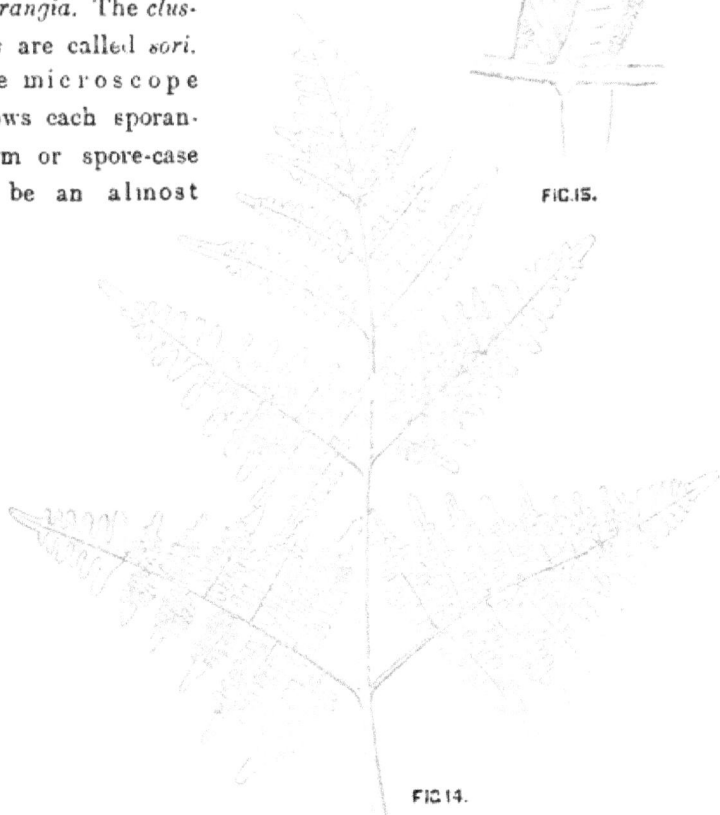

FIC 14.

globular one-celled body with a stalk attached to it, and encircled by a jointed elastic ring.   When the spore-case ripens,
the ring breaks at some point, and its elasticity then enables it
to burst open the spore-case, which then discharges its spores.

Fig. 11 shows a very much magnified sporangium whose ring has broken and ruptured the spore-case. Observe the veining of the lobes of the leaf-blade. You will see that the veins do not form a net-work. They are merely forked, and as they are not netted they are said to be *free*. The *sori* or *fruit dots* are formed at the ends of the forking veinlets (Fig. 12). The leaf of the fern, therefore, as it bears the fruit in addition to performing the ordinary functions of a leaf, is entitled to a special name. It will be spoken of as the *frond*. The petiole also will be called the *stipe* and its continuation through the frond the *rhachis*. Fig. 13 shows the peculiar way in which the frond is rolled up in the bud. Such vernation is said to be *circinate*.

Fig. 14 shows a portion of the frond of the Common Brake (Pteris aquilina). Here the frond is several times compound. The first or largest divisions to the right and left are called *pinnæ*. The secondary divisions (or first divisions of the pinnæ) are the *pinnules*. The stem, as in the Polypod, and in fact in all our ferns which have a stem at all, is a rootstock or rhizome. But here we miss the fruit-dots or sori, so conspicuous in our first example. In this case it will be found that there is a *continuous line of sporangia around the margin* of every one of the pinnules of the frond, and that the edge of the pinnule is reflexed *so as to cover the line of spore-cases*. Fig. 15 is a very much magnified view of one of the lobes of a pinnule, with the edge rolled back to show the sporangia. Some of the sporangia are removed to show a line which runs across the ends of the forking veins. To this the sporangia are attached. The veins, it will be seen, do not form a net-work, and so are free, as in Polypod. Observe, then, that in Polypod the sori are not covered, whilst in Pteris the opposite is the case. The covering of the fruit-dots is technically known as the *indusium*. The individual spore-cases are alike in both plants. (Fig. 11.)

Fig. 16 shows a frond of one of our commonest Shield-ferns (Aspidium acrostichoides). It is simply pinnate. The stipe is thickly beset with rusty-looking, chaff-like scales. The veins are free, as before. The *sori* or *fruit-dots* are on the back of the upper pinnæ, but they are neither collected into naked clusters, as in Polypod, nor are they covered by the edge of the frond as in the Brake. Here each cluster has an *indusium* of its own. The indusium is round, and attached to the frond by its depressed

centre (peltate). Fig. 17 shows an enlarged portion of a pinna with the sporangia escaping from beneath the indusium. From one forking vein the sporangia are stripped off to show where they have been attached. The separate sporangia discharge their spores in the manner represented in Fig. 11.

In some ferns the fruit-dots are elongated instead of being round, and the indusium is attached to the frond by *one edge* only, being free on the other. Sometimes two long fruit-dots will be found side by side, the free edges of the indusia being towards each other,

FIG. 17.

so that there is the *appearance* of one long fruit-dot with an indusium split down the centre.

Fig. 18 represents a frond of a very common swamp fern, Onoclea sensibilis, or sensitive fern. It is deeply pinnatifid, and on one of the lobes the veining is represented. Here the veins are *not* free, but as they form a network they are said to be *reticulated*. You will look in vain on this frond for fruit-dots, but beside it grows another, very different in appearance,—so different that you will

FIG. 18.

hardly believe it to be a frond at all. It is shown in Fig. 19. It is twice pinnate, the pinnules being little globular bodies, one of which, much magnified, is shown in Fig. 20. You may open out one of these little globes, and then you will have something like what is shown in an enlarged form in Fig. 21. It now looks more like a pinnule than when it was rolled up, and it now also displays the fruit-dots on the veins inside. Here, then, we have

FIG. 18.

FIG. 19.

FIG. 20.

FIG 21.

two kinds of frond. That bearing the fruit-dots we _ _ fertile_ frond, and the other we shall call the _sterile_ one. You must not look upon the pinnule in which the sori are wrapped up as an indusium. Sori which are wrapped up in this way have an indusium of their own besides, but in this plant it is so obscure as to be very difficult to observe.

The spore-cases burst open by means of an elastic ring as before.

Fig. 22 represents one of the Moonworts (Botrychium Virginicum), very common in our rich woods everywhere. Here we have a single frond, but made up manifestly of two distinct portions, the lower sterile, and the upper fertile. Both portions are thrice pinnate. The ultimate divisions of the fertile segment are little globular bodies, but you cannot unroll them as in the case of the Onoclea. Fig. 23 shows a couple of them greatly enlarged. There is a slit across the middle of each, and one of the slits is partially open, disclosing

FIC.23.

the *spores* inside. Each little globe is, in fact, a *spore-case* or *sporangium*. So that here we have something quite different from what we have so far met with. Up to this point we have found the sporangia collected into dots or lines or clusters of some sort. In the Moonwort the sporangia are separate and naked, and instead of

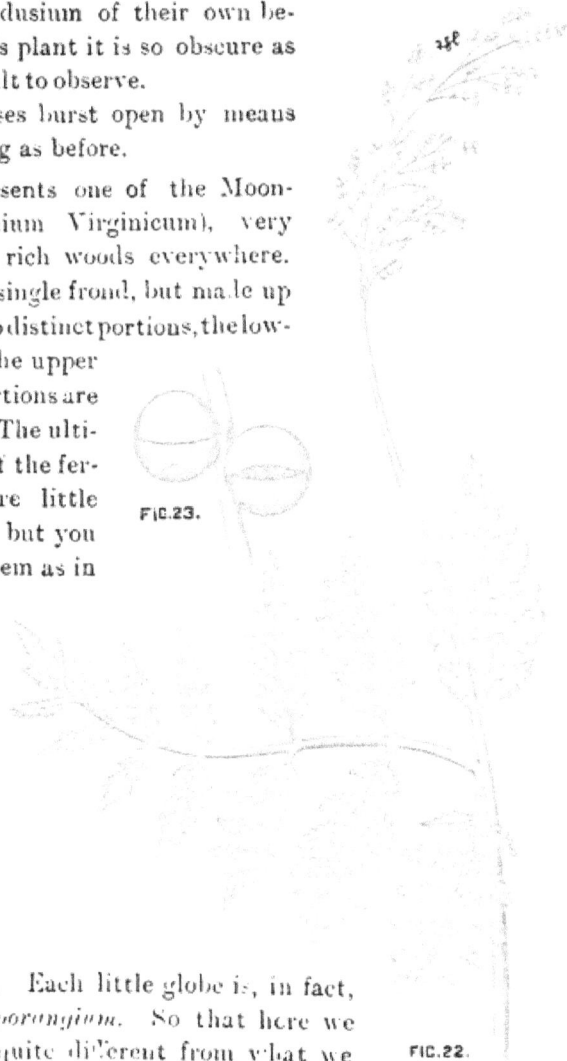

FIC.22.

bursting through the action of an elastic ring, they open by a horizontal slit and discharge their spores. In other ferns, as the Osmunda, the sporangia are somewhat similar, but burst open by a *vertical* instead of a horizontal slit.

Observe that the frond of Botrychium is *not circinate* in the bud. It is the only exception we have.

We shall now proceed to describe the commonly occurring representatives of the Fern Family.

ORDER CI. **FILICES.** (FERN FAMILY.)

Flowerless plants with distinct leaves known as *fronds*, these circinate in the bud, except in one sub-order, and bearing on the under surface or margin the clustered or separate sporangia or spore-cases.

Synopsis of the **Genera.**

SUBORDER I. POLYPODIA'CEÆ. (THE TRUE FERNS.)

Sporangia collected into various kinds of clusters called **sori.** Each sporangium pedicelled and encircled by an elastic-jointed ring, by the breaking of which the sporangium is burst and the spores discharged. Sori sometimes covered by an *indusium.*

1. **Polypo'dium.** Fruit-dots on the back of the frond near the ends of the veins. *No indusium.* Veins free. (See Fig. 10.)

2. **Adian'tum.** Fruit-dots *marginal*, the edge of the frond being reflexed so as to form an indusium. Midrib of *the pinnules* close to the lower edge or altogether wanting. Stipe *black* and *shining.* All the pinnules distinct and generally minutely stalked. Veins free.

3. **Pte'ris.** Fruit-dots *marginal.* Indusium formed by the reflexed edge of the frond. Midrib of the pinnules in the centre and prominent. Veins free. Stipe light-coloured. See Fig. 14.

4. **Pellæ'a.** Fruit-dots *marginal*, covered by a broad indusium, formed by the reflexed margin of the frond. Small ferns (3-6 inches high) with once or twice pinnate fronds, the fertile ones very much like the sterile, but with narrower divisions. Stipe brown and shining, darker at the base.

6. **Asple'nium.** Fruit-dots elongated on veins on the back of the pinnules, *but only on the upper side of the vein. Indusium attached to the vein by one edge*, the other edge free. Veins free.

6. **Scolopen'drium.** Fruit-dots elongated, *occurring in pairs on contiguous veinlets,* the free edges of the two indusia facing each other, so that the sori appear to be single, with an indusium split down the centre. Veins free. *Frond simple, ribbon-shaped,* about an inch broad, generally wavy-margined.

7. **Camptoso'rus.** Fruit-dots elongated, these near the base of the midrib double, as in Scolopendrium; others single, as in Asplenium. *Fronds simple,* ¼ or ¾ of an inch wide at the heart-shaped base, and tapering into a long and narrow point; growing in tufts on limestone rocks, and commonly rooting at the tip of the frond, like a runner. Veins reticulated.

8. **Phegop'teris.** Fruit-dots roundish, on the back (not at the apex) of the veinlet, rather small. *Indusium obsolete or none.* Veins free. *Fronds triangular in outline,* in one species twice-pinnatifid, with a winged rhachis, and in the other in three p. tioled spreading divisions, the divisions once or twice pinnate.

9. **Aspid'ium.** Fruit-dots round. Indusium evident, flat, orbicular or kidney-shaped, fixed by the centre, opening all round the margin. Veins free. Generally rather large ferns, from once to thrice pinnate. (See Fig. 16.)

10. **Cystop'teris.** Fruit-dots round. Indusium not depressed in the centre, but rather raised, attached to the frond not by the centre but by the edge partly under the fruit-dot, and generally breaking away on the side towards the apex of the pinnule, and becoming reflexed as the sporangia ripen. Fronds slender and delicate, twice or thrice pinnate.

11. **Struthiop'teris.** Fertile frond much contracted and altogether unlike the sterile ones, the latter very large and growing in a cluster with the shorter fertile one in the centre. Rootstock very thick and scaly. Fertile fronds simply pinnate, the margins of the pinnæ rolled backwards so as to form a hollow tube containing the crowded sporangia. Very common in low grounds.

12. **Onocle'a.** Fertile and sterile fronds unlike. (See Figs. 18, 19, 20 and 21 and accompanying description.)

SUBORDER II.   OSMUNDA'CEÆ.

Sporangia naked, globular, pedicelled, *reticulated,* opening by a vertical slit.

13. **Osmun'da.** Fertile fronds or *fertile portions* of the frond much contracted, bearing naked sporangia, which are globular, short-pedicelled, and opening by a vertical slit to discharge the spores. Frond tall and upright, once or twice pinnate, from thick rootstocks.

Suborder III. OPHIOGLOSSA'CEÆ.

Sporangia naked, *not reticulated*, opening by a horizontal slit. *Fronds not circinate in the bud.*

14. **Botrych'ium.** Sporangia in compound spikes, distinct, opening by a horizontal slit. Sterile part of the frond compound. Veins free. (See Figs. 22 and 23.)

1. POLYPO'DIUM. Polypod.

P. vulga're. Fronds evergreen, 4–10 inches long, deeply pinnatifi 1, the lobes obtuse and obscurely toothed. Sori large.-- Common or shady rocks.

2. ADIANTUM. Maidenhair.

A. peda'tum. Stipe upright, black and shining. The frond forked at the top of the stipe, the two branches of the fork recurved, and each bearing on its inner side several slender spreading divisions, the latter with numerous thin pinnatifid pinnules which look like the *halves* of pinnules, owing to the midrib being close to the lower edge. Upper margin of the pinnules cleft.—Common in rich woods.

3. PTE'RIS. Brake. Bracken.

P. aquili'na. Stipe stout and erect. Frond large and divided into 3 large spreading divisions at the summit of the stipe, the branches twice-pinnate, the pinnules margined all round with the indusium. Common in thickets and on dry hill-sides.

4. PELLÆ'A. Cliff-Brake.

P. gra'cilis. Fronds 3–6 inches high, slender, of few pinnæ, the lower ones once or twice pinnatifid into 3–5 divisions, those of the fertile fronds narrower than those of the sterile ones.— Shady limestone rocks ; not common.

5. ASPLE'NIUM. Spleenwort.

1. **A.** Trichom'anes. **A very** delicate little fern growing in tufts on shaded cliffs. Fronds 3–6 inches long, linear in outline

pinnate, the little pinnæ oval and unequal-sided, about ½ of an inch long. The stipes thread-like, purplish-brown and shining. This species is evergreen.

2. **A. thelypteroides.** Fronds 2-3 feet high, *pinnate,* the pinnæ linear-lanceolate in outline, 3–5 inches long, *deeply pinnatifid,* each of the crowded lobes bearing 3-6 pairs of oblong fruit-dots.—Rich woods.

3. **A.** angustifolium. Fronds simply pinnate, somewhat resembling Aspidium acrostichoides, *but very smooth and thin,* and larger. Pinnæ crenulate, short-stalked. Fruit-dots linear, crowded.—Rich woods ; not common.

4. **A.** Filix-fœ'mina. Fronds 1-3 feet high, broadly lanceolate in outline, *twice pinnate,* the pinnæ lanceolate in outline, and the pinnules confluent by a narrow margin on the rhachis of the pinna, doubly serrate. *Indusium* curved, often shaped something like a horse-shoe, *owing to its crossing the vein and becoming attached to both sides of it.*—Rich woods.

6. SCOLOPENDRIUM. HART'S TONGUE.

S. Vulga're. Frond simple, bright green, a foot or more in length, and an inch or more in width.—Shaded ravines and limestone cliffs ; not very common.

7. CAMPTOSORUS. WALKING-LEAF.

C. rhizophyllus. A curious little fern, growing in tufts on shaded limestone rocks. Frond simple, with a very long, narrow point.—Not very common.

8. **PHEGOP'TERIS.** BEECH-FERN.

1. P. polypodioides. Fronds triangular, longer than broad, 4-6 inches long, hairy on the veins, twice pinnatifid, *the rhachis winged.* The pinnæ sessile, linear-lanceolate in outline, *the lowest pair deflexed and standing forwards.* Fruit-dots small and all near the margin. Stipes rather larger than the fronds, from a slender, creeping rootstalk.—Apparently not common, but growing in rich woods near Barrie, Ont.

2. **P. Dryop'teris.** Fronds broadly triangular in outline, primarily divided into 3 triangular spreading petioled divisions, smooth, the three divisions once or twice pinnate. Fronds from 4 to 6 inches wide. Fruit-dots near the margin.—Rich woods; common. Whole plant delicate, and light green in colour.

9. ASPIDIUM. SHIELD-FERN. WOOD-FERN.

* *Stipes not chaffy.*

1. **A. thelyp'teris.** Fronds tall and narrow, lanceolate in outline, pinnate, the pinnæ deeply pinnatifid, nearly at right angles to the rhachis, linear-lanceolate in outline, *the margins of the lobes strongly revolute in fruit.* Stipe over a foot long, and usually longer than the frond.—Common in low, wet places.

2. **A. Noveboracen'se.** *Fronds much lighter in colour than the preceding, tapering* towards both ends, pinnate, the pinnæ deeply pinnatifid, much closer together than in No. 1, and not at right angles with the rhachis. Veins simple. Lower pinnæ short and deflexed.—Swamps.

* * *Stipes Chaffy.*

3. **A. spinulo'sum.** Stipes *slightly* chaffy or scaly. Fronds large, ovate-lanceolate in outline, *twice pinnate*, the pinnules deeply pinnatifid *(nearly pinnate)*, and spiny-toothed. Pinnæ triangular-lanceolate in outline. The variety **intermedium**, which is the commonest in Canadian woods, has the few scales of the stipe pale brown with a darker centre, and *the lower pinnæ unequal-sided.*—Rich woods, everywhere.

4. **A. crista'tum.** Stipes chaffy with broad scales. Fronds large, linear-lanceolate in outline, once pinnate, the pinnæ deeply pinnatifid, the upper ones triangular-lanceolate in outline, the lower considerably broader, the lobes cut-toothed. Fruit-dots large and conspicuous, *half-way between the midrib of the lobe and the margin.*—Swamps.

5. **A. Goldia'num.** A fine fern, the large fronds growing in a circular cluster from a chaffy rootstock. Frond ovate or

ovate-oblong in outline, once pinnate, the pinnæ deeply pinnatifid, 6-9 inches long, broadest in the middle, the lobes slightly
scythe-shaped, finely serrate.   Fruit-dots large, near the midrib
of the lobe.—Rich moist woods.

6.   A. marginale.   Stipes very chaffy at the base.   Fronds
ovate-oblong in outline, twice pinnate, the pinnæ lanceolate in
outline, broadest above the base.   Pinnules crenate-margined.
Fruit-dots large, close to the margin.—Rich woods, mostly on hill-
sides.

7.   A. acrostichoïdes. (See Figs. 16 and 17 and accompanying description.)—Rich woods, everywhere.

8.   A. Lonchitis.   Not unlike No. 7, but the fronds are
narrower and longer, more rigid and with hardly any stipe.   Pinnæ densely spinulose-toothed.—Apparently not common, but
plentiful in rocky woods west of Collingwood, Ont.

10.   CYSTOPTERIS.   Bladder Fern.

1.   C. bulbifera.   Frond large (1-2 feet), narrow and very
delicate, twice pinnate, the pinnæ nearly at right angles to the
rhachis.   Rhachis and pinnæ often with bulblets beneath.   Pinnules toothed.—Shady, moist ravines.

2.   C. fragilis.   Frond only 4-8 inches long, with a stipe of
the same length, twice or thrice pinnate.   Rhachis winged.—
Shady cliffs.

11.   STRUTHIOPTERIS.   Ostrich Fern.

S. Germanica. Sterile fronds with the lower pinnæ gradually
much shorter than the upper ones.   Pinnæ deeply pinnatifid.—
Common in low, wet grounds along streams.

12.   ONOCLEA.   Sensitive Fern.

O. sensibilis.   (See Figs. 18, 19, 20 and 21, and accompanying
description.)—Common in wet grounds along streams.

13. OSMUNDA. FLOWERING FERN.

1. **O. rega'lis.** (FLOWERING FERN.) Fronds twice pinnate, *fertile at the top*, very smooth, pale green. Sterile pinnules oblong-oval, finely serrated towards the apex, 1-2 inches long. either sessile or short-stalked, usually oblique and truncate at the base.—Swamps, along streams and lake-margins.

2. **O. Claytonia'na.** Fronds large, once pinnate, pale green, densely white-woolly when unfolding from the bud, *with fertile pinnæ among the sterile ones.* Pinnæ deeply pinnatifid, the lobes entire.—Low grounds.

3. **O. cinnamo'mea.** (CINNAMON FERN.) *Fertile fronds distinct from the sterile ones*, contracted, twice pinnate, covered with cinnamon-coloured sporangia. Sterile fronds rusty-woolly when young, smooth afterwards, once pinnate, the pinnæ deeply pinnatifid. The long, sterile fronds in a cluster, with the fertile ones in the centre.—Low grounds.

14. BOTRYCHIUM. MOONWORT.

1. **B. Virgin'icum.** (See Figs. 22 and 23 and accompanying description.)—Rich woods, everywhere.

2. **B. lunarioi'des** is occasionally found. It is easily distinguished from No. 1 by the sterile portion of the frond being long-petioled instead of sessile.

ORDER CII. EQUISETA'CEÆ. (HORSETAIL FAMILY.)

We shall confine ourselves to the description of a single species of the genus

EQUISE'TUM, THE ONLY GENUS OF THE ORDER.

Fig. 24 is a view of the fertile stem of Equisetum arvense, the Common Horsetail, of about the natural size. It may be observed early in spring almost anywhere in moist sandy or gravelly soil. It is of a pale brown colour, and in place of leaves there is at each joint a sheath split into several teeth. At the summit of the stem is a sort of conical catkin, made up of a large number of six-sided bodies, each attached to the stem by a short pedicel. Each of these six-sided bodies turns out, on examination, to be made up of six or seven sporangia or spore-cases, which open down their inner margins to discharge their spores. Figs. 25 and 26 are enlarged outer and inner views of one of them. The spores themselves are of a similar nature to those of the Ferns, and reproduction is carried on in the same manner; but each spore of the Horsetail is furnished with four minute tentacles which closely envelope it when moist, and uncoil themselves when dry. The use of these tentacles is not known.

The fertile stems will have almost withered away by the time the sterile ones appear. These latter are of the same thickness as the fertile ones, but they are very much taller and are green in colour. Observe, also, the grooving of the sterile stem, and the whorls of 4-angled branches produced at the nodes.

FIG. 25.

FIG. 26.

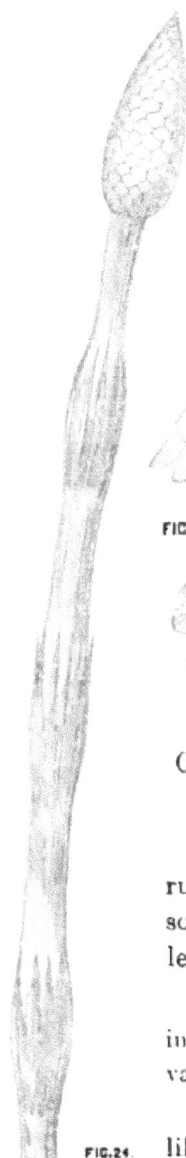

ORDER CIII. LYCOPODIA'CEÆ. (CLUB-MOSS FAMILY.)

Chiefly moss-like plants; often with long running and branching stems, the sporangia solitary in the axils of the mostly awl-shaped leaves.

1. **Lycopo'dium.** *Spore-cases of one kind only,* in the axils of the upper awl-shaped leaves; 2-valved, kidney-shaped. Chiefly evergreen plants.

2. **Selaginel'la.** *Spore-cases of two kinds;* one like those of Lycopodium, containing very minute

FIG. 24.

spores, the others 3-4 valved, and containing a few large spores.
The two sorts intermingled, or the latter in the lower axils of the
spike.   Little moss-like tufted evergreens.

1.   LYCOPO′DIUM.   Club Moss.

1. **L. luci′dulum.** Stems 4-8 inches long, tufted, 2 or 3
times forking.   The leaves forming the spike not different from
the others on the stem ; all spreading or reflexed, sharp-pointed,
serrulate, dark green and shining.—Cold, moist woods.

2. **L. anno′tinum.** Stems creeping, 1-4 feet long. Branches
4-9 inches high, once or twice forked.   Spike sessile, the leaves
of it yellowish and scale-like, ovate or heart-shaped, the others
spreading or reflexed, rigid, pointed, nearly entire, pale green.—
Cold woods.

3. **L. dendroi′deum.** (Ground Pine.)   Rootstock creeping
underground, nearly leafless.   Stems much resembling little
hemlocks, 6-9 inches high ; numerous spreading branches with
shining lanceolate entire leaves. Spikes nearly as in No. 2, 4-10
on each plant.—Moist woods.

4. **L. clava′tum.** (Club Moss.)   Stem creeping or running
extensively.   *Spikes mostly in pairs*, raised on a slender peduncle
(4-6 inches long). Leaves linear, awl-shaped, *bristle-tipped*.—
Dry woods.

5. **L. complana′tum.** Stem creeping extensively. *Branches
flattened*, forking above.   Leaves awl-shaped, small.—Dry
woods : mostly with evergreens. ●

2.   SELAGINEL′LA.

S. Rupes′tris.   A little moss-like evergreen, growing on exposed rocks in dense tufts 1-3 inches high.   Leaves awl-shaped,
with a grooved keel, and tipped with a bristle. Spikes 4-cornered.

INDEX.

The names of the Orders, Classes, and Divisions, are in large capitals; those of the Sub-orders in small capitals. The names of Genera, as well as popular names and synonyms, are in ordinary type.